Geographies of City Science

Science and Culture in the Nineteenth Century

Bernard Lightman, Editor

Geographies of City Science

Urban Life and Origin Debates in Late Victorian Dublin

Tanya O'Sullivan

University of Pittsburgh Press

Published by the University of Pittsburgh Press, Pittsburgh, Pa., 15260

Manufactured in the United States of America
Printed on acid-free paper
10 9 8 7 6 5 4 3 2 1

Cataloging-in-Publication data is available from the Library of Congress

ISBN 13: 978-0-8229-4575-8
ISBN 10: 0-8229-4575-4

Cover art: The Applied Mechanics Laboratory at the Royal College of Science, Dublin, ca. 1920. Reproduced by kind permission of UCD Archives UCDA RCSI/248.

Cover design: Melissa Dias-Mandoly

For Denis, Carmel, Mike, Andrew, and Kirsty

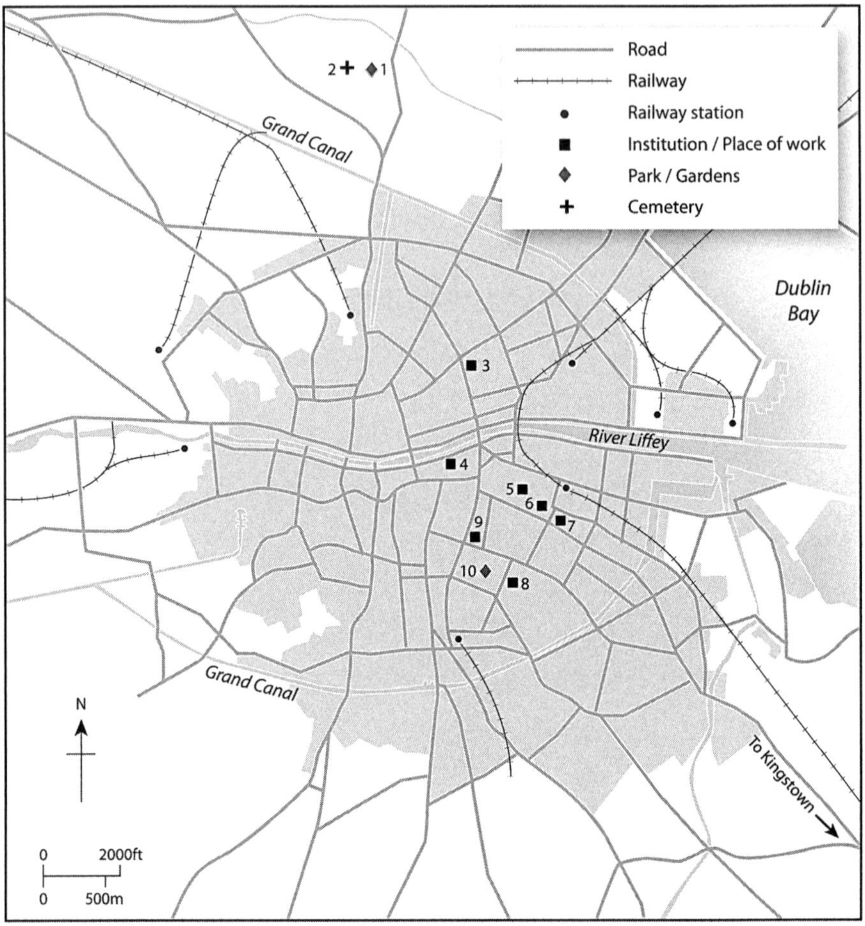

Frontispiece. Key

1. Botanic Gardens
2. Glasnevin Cemetery
3. Gaelic League Headquarters, No. 24 Upper O'Connell Street
4. The Catholic University Medical School, No. 4 St. Cecilia Street
5. Trinity College Dublin, School of Natural Philosophy
6. Trinity College Dublin, School of Anatomy and Anthropometric Laboratory
7. Medical Practice of Dr. George Sigerson, No. 3 Clare Street
8. The Royal College of Science for Ireland, No. 51 St. Stephen's Green
9. The National Literary Society Meeting Rooms, No. 6 St. Stephen's Green
10. St. Stephen's Green

Contents

Acknowledgments

This book has been over ten years in the making. During that time, many archives and libraries have provided me with access to their collections and permission to reproduce their material. I am very grateful for the kind assistance of staff at the McClay Library, Queen's University Belfast; Union Theological College, Belfast; James Joyce Library, University College Dublin; Department of History and Archives, University College Dublin; Royal Dublin Society Library, Dublin; Royal Irish Academy Library, Dublin; National Library of Ireland, Dublin; Manuscripts and Archives Research Library, Trinity College Dublin; National Botanic Gardens Library, Glasnevin; the School of Anatomy, Trinity College, Dublin; Centre For Research Collections, Edinburgh University, Edinburgh; and Special Collections, Cambridge University Library, Cambridge.

I am especially grateful to Stephen Gregory, Mary Kelleher, Philomena MacAteer, Eugene Roche and Alex Caccamo for their help with locating archival material. Kate Manning, Meadhbh Murphy, and Orna Summerville at University College Dublin Archives in Belfield, Chris Swift, Glen O' Hara, and Berni Metcalfe at the National Library of Ireland, and Sharon Sutton at Trinity's Digital Resources Imaging Service, all provided invaluable assistance later on in the project. Many thanks are due also to Gill Alexander for her work on the location map.

I owe an immense debt of gratitude to David Livingstone and Diarmid Finnegan for introducing me to the geographies of science, and for their guidance, encouragement and wise criticism along the way. I have benefited enormously from conversations with, and encouragement from, colleagues and friends who also share an enthusiasm for both the geographies of science and the history of science in Ireland. For their insightful comments, advice and suggestions at various stages of the project, I thank Nicolaas Rupke, Nuala Johnson, Jonathan Wright, and Andrew Holmes. In Belfast, I learned a great deal by attending meetings of the 'Cultures of Science and Religion' reading group at Queen's University. Thanks to Ciaran Toal, Mark Wood, Duncan Taylor, Nuala Johnson, Max Meulendjiks, Stewart Mathieson, Emma Swain, Rory Mawhinney, Diarmid Finnegan and David Livingstone, for finding, sharing, and discussing some fascinating papers over the years. Since its foundation in 2014 in Dublin, the History of Science, Technology and Medicine Network (HSTM) Ireland has provided a much needed platform for the presentation and discussion of ideas on science and culture in a historical context. I am grateful for the opportunities it has provided me to share my research. Thanks to colleagues Juliana Adelman, Adrian Kirwan, and Elizabethanne Boran for their vision and persistence in keeping it afloat, and to all the other HSTM members who strive to keep the subject alive in Ireland where it is not officially taught at third level.

I am greatly indebted to Oliver Hochadel and Agustí Nieto-Galan for not only reading and commenting on earlier portions of the manuscript, but for seeing a place for it in their recent edited volume *Urban Histories of Science*. Thanks to both for sharing their many insights on the scientific culture of cities. A special word of thanks goes to Bernard Lightman who read and gave invaluable feedback on the entire manuscript, for which I am immensely grateful. I am also indebted to two anonymous referees who evaluated the manuscript on behalf of the University of Pittsburgh Press. I am particularly thankful to Abby Collier and all the staff at the press for getting the manuscript into production and print.

Family and friends have supported this project in a variety of invaluable ways and this book would not have been realized without their help and encouragement. Aonraoi de Paor and Ray Bates kindly passed on to me many history of science volumes from their own collections. I am deeply grateful to them. Finally, my greatest debt is to Denis, Carmel, Mike, Andrew, and Kirsty, without whose enduring support and encouragement the book would not have been possible.

Geographies of City Science

Introduction

> The origin of life is a mythical concept, not in the sense of being untrue but rather in stirring a deep sense of mystery, even scientists need to narrate, to integrate their observations into origins stories.
>
> –**Lynn Margulis**

As Lynn Margulis observes, discussions around the fundamental questions of our origins, from the dawn of the universe to the genesis of social networks are perennial ones that resonate across many disciplines. Popular science books from Stephen Hawking's *A Brief History of Time* to Martin Rees's *Our Cosmic Habitat* trace stories that start from swirling clouds of cosmic dust and end with the emergence of mind and human culture. Rees remarks, "We now envision our earth in an evolutionary context stretching back before the birth of our solar system."[1] This human preoccupation with an origins narrative has also continued to fascinate the minds of historians and historians of science.[2] James Secord, for instance, queries how the evolutionary idea gained a pivotal role in the public arena, and he concludes that it had little to do with Darwinian biology or Big Bang astronomy and advocates instead a nineteenth-century event as the critical turning point.[3] This pivotal moment he argues occurred in the first half of the nineteenth century with the anonymous publication of Robert Chambers's *Vestiges of The Natural History of Creation* in 1844. This publication and its sequel *Explanations* in 1845 introduced a developmental cosmology based on natural law to the English-speaking world. Chambers's aim, according to Secord, was "the universal application of a simple principle of development, to do what Newton had done with gravitation in the *Principia*."[4] Taking Secord's assessment into account we can see how the message and widespread circulation of Robert Chambers's book paved the way for the creation of a progressive past for an industrial imperial society. The "nebula to man" cosmology thus became the grand narrative of development in which the "how" and "when" questions of origins and futures could be explained to a science-hungry Victorian public during a time when specialized experts increasingly assumed authority for the making of knowledge.

This book on Dublin engages with the familiar evolutionary trope but in a different way. At every stage, recourse to recent themes in historical geography dominate, since the aim here is not to emphasize the many heroic attempts to answer fundamental questions of origins, but rather to explore their negotiation in a particular place and time. *Geographies of City Science* contends that scientific debate is significantly shaped by the places in which scientists work and that examining Victorian origins discourse from the perspective of "where" allows for a more useful discussion of "how" such knowledge was constructed, communicated, and contested. Being a city on the cusp of change, from a seat of empire (albeit peripheral), to a national capital, late nineteenth-century Dublin offers a rare opportunity to observe the dynamic interplay of urban space and contemporary science. And, as we shall see, its scholars aligned themselves with vastly different political, social, and cultural mores.

Methodologically, the study takes the form of local Dublin case studies that combine archival research with broader thematic reflection on origins. Eight intellectual figures have been selected based on their contributions to the nineteenth-century science of "origins" broadly conceived. Together they embody a range of cultural, political, religious, and disciplinary backgrounds reflecting the diverse and changing character of fin-de-siècle Dublin. The stories congregate around contemporary debates on the origins of matter, life, humans, and language and aim to seek out some of the numerous ways in which scientific meaning relating to these questions was established and mobilized. Traditional attempts at explaining Victorian scientific controversies in Dublin have used the model of religious division. Each vignette here also explores the lives of two practitioners from the main religious/political traditions in Dublin (either Protestant/Unionist or Catholic/Nationalist) but, this time, to underscore the thesis that any variation in the engagement with science had more to do with the aggregate life-spaces of an individual, of which religious divisions were merely a part. Rather than valorizing the scientific contributions of eight city scholars, the book seeks to inform our understanding of "origins" issues as a coproduction of scientific lives and urban space.

The theoretical approach taken here follows the geographical "turn" to space in which studies of the making and mobilizing of science are scrutinized by sociospatial analysis.[5] In different locations, at different times, in different circumstances, and at different scales, space has been shown to be an active ingredient in social and cultural life that has made its mark on science in different ways.[6] With this in mind, and taking "origins" as its connecting theme, the book asks two major questions: How might we better understand what is meant by a "scientific life"? And how can these findings add to our knowledge of the way "origins" science was interpreted and carried out in the city? These questions necessitate two introductory discussions here: the first on the reenvisioning of scientific biography and the second on the role of space and place in city science.

Scientific Biography Recast

Can "a life as it is told" capture the essence of "a life as it is lived"?[7] If we accept that space is an active ingredient in social and cultural life that affects science in different ways—and as a consequence we view scientific theories as forms of work in the context of everyday lives—where does that leave the traditional genre of scientific biography? No matter how strong their emphasis on "social explanation" or "context," biographies tend to keep the individual at the center of readers' understanding. If what once made sense as the "Darwinian revolution" for instance, can now be seen according to Secord, as an episode in the industrialization and transformation of reading audiences, new questions need to be asked about the role of biography as a historiographical tool. How, for example, has biography responded to the geographical turn in the history of science and in what ways can it lead us to a better understanding of scientific lives?[8] To answer these questions, the following few pages sketch out some reemerging trends relating to the biographical perspective in the history of science and offer a new life-space perspective that lays the ground for a fresh historiographical approach to science in Dublin.

Biography is one of the principal narrative modes of the history of science, and reflection on its role in scholarship is also an ongoing tradition of the discipline. The history of science, positioned between the humanities and the sciences, has found the biography a persuasive means of fulfilling its mission of uniting two cultural realms: idiosyncratic personhood and public scientific knowledge. In the words of the historian of science Mott T. Green, "To write a scientific biography is to document the heroic transformation of inner experience into public knowledge"; but, he warns, the continued use of conventions such as the Weberian ideal types, Bildungsroman, and folkloric tales of the "hero's quest," has exerted a distorting influence on the image of how most science gets done.[9] Green contends that science—like all forms of human activity—involves ordinary people, doing ordinary things in ordinary ways, and so he believes that biography cannot by its very nature provide a picture of how science works most of the time.

From the inception of the *New Dictionary of National Biography*, some historians voiced misgivings about the enterprise and they expressed objections to perpetuating a tradition of writing the history of science as the biographies of great men and great ideas.[10] However, some groundbreaking biographies have emerged from the "life and times" genre since the 1970s. Adrian Desmond and James Moore's biography of Darwin provided, in Moore's words, "a focus for a biographical genre of historical writing that may yet prove the most effective way of informing the widest audience about the politics of scientific practice and the cultural formation of natural knowledge." Janet Browne writes of her subject, Darwin, that his is the story, not just of a life, but of the era.[11]

Another story of the era—which has languished for a century in the shadow of Darwin's—was that of Richard Owen. Through his revival of the life and work of Owen, Nicolaas Rupke has succeeded in bringing to the reader's attention, the important role of *place* and *practice* in the production and reception of nineteenth-century scientific knowledge.[12] In *Energy and Empire: A Biographical Study of Lord Kelvin*, Crosbie Smith and M. Norton Wise used the biographical device to demonstrate how Kelvin's science emerged from his culture, and how he drew on the conceptual and material resources of his industrial milieu to arrive at physical theories. Add to these Richard Westfall's Newton study, David Cassidy's Heisenberg biography, Geoffrey Cantor's study of Faraday, Frederic Holmes's volume on Hans Krebs and we begin to see a conscious move away from biography in terms of great men and their ideas, to inquiries into the cultural resources of scientific theory and the social construction of scientific knowledge.[13]

These endeavors heralded a change of direction certainly, but biography still falls within the traditional confines of the genre according to some history of science commentators. Christopher Chilvers points out that the relationship between the history of science and "scientific biography" is a far more complex configuration, counterintuitive even, than is often recognized, and argues that consideration should be given to genres such as tragedy. This he illustrates with his portrayal of the life and early demise of the Soviet physicist and Marxist historian of science, Boris Hessan. He maintains that to adequately convey the contexts of a modern scientific life, the structure of the "scientific biography" requires adaptability and a subtlety that can often be overlooked.[14]

Thomas Söderqvist also opens up the recasting of science biography and suggests it should not be seen primarily as yet another means for disclosing the contextual and socially constructed nature of science, but a genre that conveys an understanding of what it means to live a life in which scientific work and rational thinking are part of an existential project and involve existential choices. He cites Ray Monk's biography of Wittgenstein as an example of modern science hagiography—in the literal, not pejorative sense of the word, more "a mode of communal self-scrutiny."[15] What distinguishes his existential biography from more traditional ancestor worship is the much greater range of lives to learn from, so that whereas in the past the exemplary principle worked in favor of tradition, today it works in favor of pluralism. This, he believes, can build on recent work around scientific practices undertaken by sociological investigations of science.

Similarly, although individuals often figure as convenient units of study in the history of science, only rarely are scientists depicted as whole persons for whom science is part of the meaning of a life.[16] Theodore Porter argues that the culture of science shapes and is shaped by the people who practice it, and that the scientist, as a human type, has a history that matters. This view of a unified narrative for science within a life—or what the anthropologist James Clifford has referred to as "the myth of personal coherence"—is questioned by the historian of science Mary Jo Nye, who offers a variety of ways to understand the lives of scientists. She claims that although biography is often used as a vehicle to analyze scientific processes and scientific culture, the most compelling scientific biographies are those that portray "the ambitions, passions, disappointments, and moral choices that characterize a scientist's life." She sees the many facets of a scientific life as a given, and as so distinct that they will not be accessible to any single audience.[17]

Other commentators have focused on the "man of science," an entity that preceded the modern (fragmented) scientist of Nye's critique. Here, although Mary Terrall underlines the very real tension between the study of individuals and the study of disciplines, institutions, ideas, and practices of the eighteenth century, hers is a conundrum that could equally be applied to any era. In describing her approach to the biography of Pierre-Louis Moreau de Maupertius she explains, "I hoped to finesse the old internal/external problem by looking closely at the place of this one man in his many contexts."[18] In a similar vein, Nicolaas Rupke's metabiography of Alexander von Humboldt has dissected many biographical accounts to reveal the diverse guises under which von Humboldt has appeared, and more recent expositions of the theme have examined the changing public image of the British "man of science" over the course of the nineteenth and early twentieth centuries.[19] The essential point highlighted by these studies is that all scientific biographers must locate their subjects in a complex landscape of institutional structures and definitions of science. Allied to these considerations, many of the authors remind us of the difficulty of bringing to readers the lingering tension that exists between the social and cooperative nature of scientific thought on the one hand, and individual insight on the other.

As a result, our attention today is drawn to new research that has carved out a middle course between scientific biography and the history of science and is more attentive to how cultural categories intersect with individual lifelines.[20] These studies do not simply point to a new defense of biography, but to the importance of personal identities and life histories in the cultural history of science. It is therefore imperative, returning to Green, that future biographical treatments of scientific lives should "situate their subjects in the institutions and networks that made their careers possible" in order to balance the distortions created by traditional genre conventions."[21]

Overlapping Terrains of Biography and Geography

This volume argues that the biographical narrative is especially suited to a spatial interrogation and understands that the lives of scientists are significantly shaped by the places in which they work. It contends that the quest to chart specifically how people negotiate their everyday experiences—a crucial element of the practices of science and its cultural location—is a fundamentally geographical one. Perhaps the most powerful and useful features of recent directions in biography therefore, are those aspects of scholarship that overlap with geographical inquiry. Human spatiality is after all, as Edward Soja contends, the product of both human agency and contextual structuring. On the one hand, our actions and thoughts shape the spaces around us, but at the same time, "the larger collectively or socially produced spaces within which we live also shape our actions and thoughts in ways in which we are only beginning to understand."[22]

Scientific biographers have already been moving toward geographical critiques by being led "into the worlds of readers, spectators, institutions, collaborations, disputes and all the other interactions that make up the life of science."[23] These alliances of geography and biography are natural marriages according to Stephen Daniels and Catherine Nash who observe that the narratives of the life path in western culture have long been plotted in an explicitly geographical way. On the other hand, recent research in biography has implications for geographers with an interest in the history of science. The greater sensitivity to "the spaces of a life" has opened up new and revealing ways for historical geographers to take the measure of a life. As David Livingstone proposes, "A greater awareness of the spaces of biography, or life geographies would enormously enrich our understanding of the mutual making of science and the scientist."[24]

The role of "scientific lives" in wider interacting networks of knowledge has been explored by the geohistorian Martin Rudwick. He claims that what is needed for a fuller understanding of the processes by which scientific knowledge is shaped are empirical studies of science in the making, which focus not on one individual scientist but on a specific scientific problem that brings together groups of individuals in an interacting network of exchange. His geographical approach to scientific lives introduces the idea of an urban "cognitive topography" and nudges closer to the concept of this book. Rudwick's mapping of the London location of key figures in the Devonian controversy shows that in the making of geological knowledge, physical location, social positioning, and cognitive authority were intricately woven. His study begins by establishing the sociological and scientific environment in which the ideas were cultivated. He then underlines the importance of major institutions such as the Geological Society of London and proposes that through the social means of argument and debate, especially among a small group of key men, scientific knowledge was shaped from the materials of the natural world and that empirical evidence constrained but did not determine the eventual outcome.[25] Rudwick's move away from the individual biography was fueled by the belief that it gave less than adequate attention to the complex web of social and cognitive interactions that bind scientists into their networks and colleagues, but the perspective that this book offers approaches from the direction of more recent attempts to acknowledge the role of *personal identities* in the production of science. This can be best understood in the context of recent work, particularly in the sociological studies of science.

The Sequestration of Experience

As Söderqvist and Porter suggest, scientists do not merely reflect their context but seek out a position within a field of possibilities. This idea resonates with the sociologist Anthony Giddens when he speaks of "the sequestration of experience" as a feature of modernity, and observes that all human beings in all cultures preserve a division between their self-identities and the performances they put on in a specific social context.[26] We are also reminded by Amélie Rorty that an individual human being may be regarded as a host of personae, each of which is "a distinct and unified agent, a locus of responsibility for a range of choices and actions."[27] Perhaps the concept most pertinent here has emerged from the scholarship of Pierre Bourdieu, who surveys the contact zone where social laws and individual minds meet, and who argues that our proper object of analysis should be this middle ground.[28] *Habitus*, as he terms it, is an invaluable tool for exploring the interdependence of human agency and social structure. Bourdieu's framing of habitus as a system of dispositions is important for this study of scientific endeavor in Dublin, because dispositions are public and observable, they enact preferences, and are therefore declarations of where one stands and what one's allegiances are. The idea of a scientific habitus has already been developed in numerous ways by researchers in the social sciences.[29] As the French sociologist Remi Lenoir explains, by studying the actors and institutions of the scientific world as well as the permeable borders between this world and the field of ideological production, "we create resources enabling us to understand the epistemological limits of the scientific construction of the world."[30]

This concept too is a profoundly geographical one. Researchers such as Nigel Thrift and Gary Bridge call on habitus as a fundamental idea in understanding what Thrift terms "knowledges of position" as forms of social power.[31] This merges well with the recent thoughts of historical geographers too. Daniels and Nash again comment: "The conceptual dispersal or decentering of the autonomous individual and unified life pursued by recent historical geographers has if anything emphasized the intersection of the geographical and biographical in overlapping domains of self and place, positionality and identity, spatiality and subjectivity."[32]

Ultimately, what has emerged from this wide range of research, is a set of studies concerned with "the scientific persona" most recently visited by Lorraine Daston and S. Otto Sibum. Their maxim that "in between the individual biography and the social institution lies the persona: a cultural identity that simultaneously shapes the individual in body and mind and creates a collective with a shared and recognizable physiognomy" is largely inspired by the anthropologist Marcel Mauss's social history of the categories of the human mind.[33] Unlike Rudwick's interacting scientific network, which is based on the mapping of individuals as historical entities, the "scientific persona" is a collective entity whose Cartesian coordinates cannot be precisely defined. However, personae are as real as biological individuals, in that "they create the possibilities of being in the human world, schooling the mind, body and soul in distinctive and indelible ways.[34]

The scientists in this study of late Victorian Dublin occupy an immense range of different city spaces in which they act differently, call on different linguistic repertoires, and project different selves. By drawing on concepts such as Bourdieu's "habitus" and Rudwick's "scientific web," the following chapters explore the multiplicity of spaces that emerge from the interaction between the Dublin society that granted significance to the "persona" and the individuals who embodied it. By scrutinizing the many constraining and enabling spaces that these individuals brought—as Unionists, Nationalists, Catholics,

Protestants, curators, medics, and others—to the scientific table, we begin to see some of the ways in which spatial approaches to biography can offer alternative possibilities in the history of science. Paying close attention to the lives and local circumstances of scientific practitioners in the late Victorian city, that is, to the ways in which knowledge relating to origins was first absorbed, then reviewed, staged, and judged for audiences, scholarly cultures of scientific meaning can be plotted and we can begin to interrogate the traditional Catholic/Protestant and Unionist/Nationalist models of scientific historiography in Dublin.

Life Space and Urban Space

This leads to the second question the book poses: How can this new spatial understanding of biography add to our knowledge of the way "origins" science was interpreted and carried out in the city? Our present understanding of the history of metropolitan science more generally owes much to the dedication in 2003 of a special volume of *Osiris* to "science and the city." Although investigations of the interactions between the city and the production of knowledge have since culminated in comprehensive publications such as *Urban Modernity in the Second Industrial Revolution*, many of these studies have tended to focus on the major European hubs of Paris, London, and Berlin.[35] This book, by foregrounding Dublin, adds to more recent attempts to enrich this historiography by examining science in less well-known urban centers. It therefore builds on the perspective embodied by the research agenda of STEP (Science and Technology on the European Periphery) and the recent edited volume *Urban Histories of Science: Making Knowledge in the City*, which spotlights other "second," "smaller," and "peripheral" European cities such as Barcelona, Budapest, Helsinki, Glasgow, Lisbon, and Naples.[36]

To some extent the way has been paved for this new approach to science in Dublin by recent history of science publications in Ireland that have critically reexamined the ways in which science has both shaped and been shaped by Irish culture.[37] Juliana Adelman, for instance, moves the emphasis away from traditional questions of what science did for Ireland by focusing more on what science meant to the Irish public. Her research explores the ways in which science became an important aspect of middle-class culture and identity in the nineteenth century. Others also looking at the national picture in nineteenth-century Ireland have highlighted the connections between voluntary societies, politics, and science as well as examining the popularization of science among provincial audiences. What excites historians about these new endeavors is, according to James Bennett, "the recognition of the cultural location of science in Ireland's past, the contested nature of science as a cultural resource and the political context for its management."[38] Past studies of science's history in Dublin have indeed been played out in terms of contested cultures; a colonial culture that was essentially British, imperialist, scientific, technological, and predominantly secular and materialist;[39] and an Irish/Celtic Nationalist counterculture that embraced spirituality, literature, mythology, and was nonscientific, and antimaterialist in essence.[40] These traditional narratives could transpose neatly onto a *stagelike* vision of Dublin on which institutional forms, scientific beliefs, and epistemic traditions varied from place to place and were differently influenced by social, cultural, and political contexts. But as authors such as Sven Dierig and colleagues contend, the city is a place in which behavior can never be fully predicted or controlled, so, Dublin as a crucial *space* for science must be looked at afresh. The spatiality of scientific culture in Ireland has indeed already been explored in relation to the reception of Darwinism in Belfast, and particular Dublin institutional spaces have served as comparators for wider debates on the spatiality of scientific knowledge.[41] But questions around place and space, territory and topography, cartography and scale have been less audible from historians of science who have in the past made Dublin the *focus* of their inquiries. To that end, "Dublin science" is here placed within the parameters of a historical critique—geographically construed—which marks a clear departure from previous historiographical approaches to history of science in Dublin.

In past studies of city science, space has tended to be viewed primarily as an architecturally built environment, shaped and reshaped over time in historical processes of urban development that give little attention to its more dynamic and generative qualities. However, since the 1980s authors with an interest in urban science have become increasingly attentive to the idea of the city as a sociospatial setting that affects the production of knowledge in various ways. Sophie Forgan for instance, has shown the way that architecture advertises the intended place of science within the culture of a city and, in a similar vein Adi Ophir and Steven Shapin have called for an examination "of the manner in which the conditions of our knowledge vary according to our placement in social and physical space." Other scholars, too, have detailed the way practices and characteristics of the urban environment have molded scientific knowledge.[42]

In casting a heuristic eye over four interconnections of science and the city, the *Osiris* volume edited by Dierig, Lachmund, and Mendelsohn drew significant attention to the interaction between science and urban everyday life. Examples of traffic between urban and scientific knowledge practices were examined suggesting a continuum rather than an essential difference between them. The authors concluded that science and the city are not merely two complex historical and cultural configurations, but are in fact two continually interacting entities. Some of these urban themes had their genesis in the work of Carl Schorske who argued that the city produces cultural and scientific innovations more because it is the scene of political and social upheaval, and thus a crucible of culture change. If, as these authors imply, city life is virtually synonymous with disturbances of the senses, mind, and body the inference that urban distraction can influence the way science is done, urges closer attention.[43] Oliver Hochadel and Agustí Nieto-Galan's detailed scholarship on science and technology in turn-of-the-century Barcelona (2016) also works on the assumption that urban space creates and facilitates knowledge production and circulation and is itself also substantially transformed by these practices.[44]

Of course spatial conceptions of Dublin have had previous airings in the literary world, Michael Begnal for instance, in his introduction to *Joyce and the City: The Significance of Place* reflects on the concept of space, and holds it responsible for generating the various levels of "cityscape" evident in James Joyce's turn-of-the-century Dublin. Here, autonomous notions of the individual and the city are eschewed in favor of a complex set of interacting spaces. The inner/outer tension between urban space and spaces of the mind, was, as Begnal explains, central to Joyce's understanding of what is meant by "the city."[45] All the literary contributors to Begnal's volume move beyond traditional definitions of the city and bring contemporary critical and theoretical perspectives to bear on the very idea of space itself. These ideas resonate with a recent emphasis in geography on the active agency of space that has characterized cultural analyses of urban life,[46] in particular, Soja's exposition of "cityspace" as a "distinctive material and symbolic context or habitat for human life." Like Joyce, Soja's work probes the intrinsic "processual or dynamic" aspects of cityspace, as well as its formal and morphological features.[47]

We only have to look at Dublin tourist guides, memoirs, histories, statistical surveys, and literature that direct their gaze at the turn-of-the-century city to see that they all "obtrude a differently coloured Dublin on us."[48] A quick review of titles detailing city experiences, from Frances Gerard's *Picturesque Dublin*, praising castle and military life, to the slums of Sean O'Casey's famous Dublin trilogy, reveals a profusion of city spaces where social class, local politics, religious creed, and cultural identity intermingle. In *Ulysses*, Joyce attempted to capture the essence of this culturally variegated Dublin life. As he tells us, "I tried to give the colour and tone of Dublin with my words: the drab, yet glistening atmosphere of Dublin, its hallucinatory vapours, its tattered confusion, the atmosphere of its bars, its social immobility."[49] Acknowledging this new concept of city space and being cognizant of the active role cities play in shaping scientific endeavor as Dierig, Lachmund, and Mendelsohn have suggested, we can begin to grasp some of the ways in which "the city" might be viewed as a fertile site for understanding the locatedness of scientific endeavor.

Placing Dublin Science

Nineteenth-century Dublin science inhabited many city places. Nicholas Whyte in his critique of the cultural background to science in Ireland proposes three particular strands of scientific endeavor.[50] The first, which he describes as "Ascendancy" science was carried out mainly in Trinity College, the Royal Dublin Society, and the Royal Irish Academy, predominantly by Irish Protestant scientists. Their metropolis was to be found mainly between Trinity, Leinster House, and Dawson Street. The second strand involved institutional or professional Irish science and the increasing influence of the South Kensington Department of Science during the nineteenth century. The employees working at institutions such as the Dublin Museum of Science and Art were mainly English and Scottish, and to a lesser extent Irish and German. The third strand involved the majority Catholic and Nationalist tradition represented by centers in Maynooth and University College Dublin. Whyte maintains that all three strands encompassed an exclusive and excluding intellectual community of its own.

His approach is in keeping with the tendency in historiographical studies of Dublin science to view its endeavors under the rather broad rubrics of "Catholicism" and "Protestantism."[51] From the point of view of this research however Whyte's breakdown of science in Dublin lends itself well to a geographical assessment by flagging the distinct physical and cultural places in which science was carried out in Dublin.

However, such "places," as Charles Withers and David Livingstone tell us, are "far from being just matters of physical location and symbolic meaning. They are also constitutive of social exchange."[52] Perceiving the laboratories of the Royal College and Trinity College, for instance, as sites where scientific knowledge was produced and collected, and viewing Dublin lecture halls, cafés, and city drawing rooms as places where information was disseminated and received in a variety of ways, allows us to appreciate the extent to which the nature of science was consequential upon the social relations at work in these places. It also enables the reemergence of less well-known scientific contributors, such as technicians, assistants, and family members who played a part in generating and communicating knowledge there. Demonstrating that

No.	Ward	Roman Catholic	%	Church of Ireland	Presbyterian	Others	Total
1	Aran Quay	29 673	84.7	4525	307	514	35 019
2	Clontarf East	2080	63.0	822	192	209	3303
3	Clontarf West	3163	55.9	1488	473	538	5662
4	Drumcondra	7394	75.6	1729	265	386	9774
5	Glasnevin	6785	64.7	2785	495	416	10 481
6	Inns Quay	21 157	90.8	1823	109	216	23 305
7	Mountjoy	23 062	88.7	2507	230	199	25 998
8	North City	7251	91.9	467	76	92	7886
9	North Dock	20 966	85.6	2573	447	520	24 506
10	Rotunda	13 867	88.8	1470	169	111	15 617
11	Fitzwilliam	8805	91.9	2534	223	842	12 404
12	Mansion House	9567	85.6	1366	103	153	11 189
13	Merchants' Quay	21 386	83.2	3104	242	974	25 706
14	New Kilmainham	7307	75.9	1973	85	265	9630
15	Royal Exchange	5226	80.0	1077	94	134	6531
16	South City	3148	81.6	554	44	111	3857
17	South Dock	12 782	84.8	1909	151	232	15 074
18	Trinity	9696	87.5	1196	83	106	11 081
19	Ushers' Quay	22 575	87.5	2765	182	271	25 793
20	Wood Quay	17 480	79.5	2690	187	1629	21 986
21	Pembroke East	8507	66.8	2957	584	689	12 737
22	Pembroke West	11 193	67.6	4498	307	559	16 557
23	Rathmines East	10 366	54.2	6607	714	1431	19 118
24	Rathmines West	10 087	53.9	6287	950	1398	18 722
25	Blackrock	6118	67.4	2371	195	396	9080
26	Dalkey	2432	68.8	903	65	136	3536
27	Killiney	1943	71.4	702	41	35	2721
28	Kingstown	11 961	69.5	4399	360	499	17 219
	TOTAL	315 977	78.1	68081	7373	13 061	404 492

FIGURE 1. Religious Affiliation of the Population of Dublin (1911). Source: Ireland. Census of Ireland 1911: Province of Leinster City of Dublin and County of Dublin Cd-6049-II and Cd-6049-I (Dublin 1912). From Christopher, "Second City."

science can be understood geographically in Dublin is thus the foundation from which new and rich stories can be built.

To begin with, we might ask how the location of these different city venues provided repertoires of meaning that facilitated communication between scholars. For example, familiarity with the customs of an institution such as the Royal College of Science, the National Museum, Trinity College, or University College Dublin would have been fundamental to decoding the messages within which communication was embedded. As Livingstone puts it, "The task is to make particular sense of particular rules in particular places."[53]

Audiences for science have geographies too, and *where* ideas are received can influence *how* they are received. Iwan Morus has argued that what mattered for an audience's appreciation of science and its epistemological status nationally was quite literally the geographic/cultural locations of the places in which they occurred. This particular arena of inquiry has given rise to new *cartographies of science*—termed "geographies of reading" by James Secord and "geographies of reception" by Nicolaas Rupke—which have opened up questions about the way texts are interpreted in different venues.[54] These ideas have enabled scholars to appreciate that the meaning any new work has for an individual reader is shaped by the other texts, theories, and practices with which they have already engaged. In other words, readers come to texts with different reading histories and the idea in transit—whether accommodated or rejected—is, as Edward Said explains, to some extent "transformed by its new uses, its new position in a new time and place."[55] In another arena, attending to the "voice of science" too has been found to require sensitivity to the geography of scientific speech. In institutional settings, the ways in which scientific theories are talked about require special care because speech spaces shape what can and cannot be said in particular venues.[56]

My investigations will therefore congregate around the "territories" inhabited by the different strands of science in Dublin to demonstrate, for example, that knowledge produced in College Green or Earlsfort Terrace mattered in terms of the way it was communicated or interpreted. As this study shows, Dublin's various epistemic traditions intermingled in city spaces in ways that influenced scientific activities carried out within their confines.[57]

One major contributor in this regard, which interlaced with the scientific and political terrain of Dublin, was religious denomination. An examination of the 1911 census returns indicates a widespread distribution of both Catholics and Protestants across the city and the peri-urban fringe. Catholics were in a substantial majority in practically all of the district electoral divisions apart from the Rathmines and Pembroke areas, which had sizable Protestant communities, predominantly Anglican, but also Presbyterian. Rathmines returned the only Unionist member of the Westminster Parliament to be elected in Ireland outside Ulster.[58] However, although at first glance Dublin's religious communities appear to be associated with distinct territories, an examination of residential segregation in the late colonial period highlights the fact that urban segregation levels in Dublin were weak compared with colonial models elsewhere. According to A. J. Christopher, Dublin's religious communities were remarkably integrated with segregation reflecting social and economic status more akin to the contemporary North American city than the colonial model.[59] Denomination was not the only religious consideration in the exchange of late Victorian scientific ideas, however; theological position with regard to scientific claims also played a part, and that allowed for much intra-denominational debate. As important theological questions were reassessed in the light of nineteenth-century developments in earth sciences and biology, many religious leaders and men of science felt especially obliged to seek to resolve the tensions and preserve a coherent worldview for their spiritual followers. Presented with this political, social, and religious suffusion in late Victorian Dublin, we begin to see how "scientific personae" emerge and develop in different spaces, and how these cityscapes shape the individuals that incarnate them and vice versa. The emphasis now will move away from Whyte's bracketing of "Nationalist," "Professional," and "Ascendancy" science and focus instead on the role of these strands as facets of a scientific life in the city.

This life-space critique adds to the growing suite of analytic categories emerging from the field of history of science today.[60] However, the more dynamic interplay of places, roles, and communicative practices that arise from studies such as this has led Lynn Nyhart to observe that the subject can now be more fittingly viewed "as a densely tangled bank of people and material things teeming with social, cultural, economic and religious life."[61] This alerts us to the gap that has opened up between the "big picture" historiographical approach of the past and the rich local stories of today, but has also led to some accusations of parochialism. Secord, for instance, warns that the lure of the local and its abundance of microhistories may just lead to "a rich array of research that adds up to less than the sum of its parts."[62] But recent research has demonstrated that the discipline has adapted and responded to such accusations by forging a middle course, which accepts that studies of circumscribed localities can provide a different way of investigating the "big questions" surrounding process and structures.

Any good historical geography of science today assumes that place can only be convincingly understood in the context of much wider forces—social, economic, political, and cultural—acting upon and through it. As Simon Nayor points out, historical geographers should always keep in mind that places function—indeed are actively constituted—in relation to many other places.[63] Robert Kohler agrees when he calls for a "generalist" account of scientific change that does not presume a positivistic understanding of science as universal by definition.[64] As this city/life study underlines, conceptualizing social relations in Bourdieuan terms gives rise to an image of society as a web of interrelated spaces. This complex dynamic ensures that "at every scale individuals inhabit locations that at once enable and constrain routine social relations."[65] Importantly, our "performances" as spatial beings take place at many different scales "from the body . . . to a whole series of more distant geographies, ranging from rooms and buildings, homes and neighbourhoods, to cities and regions, states and nations."[66]

In the chapters that follow, the concept is addressed at both the micro and meso levels and along the spectrum of scales from particular Dublin institutional debates through to the ways in which science becomes entangled with national and international concerns. The investigative parameters of chapter 1 sweep from the nineteenth-century city center to its suburban coastline at Kingstown, and encompass in its reach the divergent mobilization of ether theories by two Dublin physicists, George Francis Fitzgerald (1851–1901) and William Fletcher Barrett (1844–1925). Fitzgerald, a professor of natural philosophy at Trinity College, was a Dublin-born Anglican, a Unionist, and member of an elite circle of Dublin scientists centered around Trinity. William Fletcher Barrett, a non-Dubliner in Dublin from a nonconformist congregational background, was based at the newly established Royal College of Science for Ireland. He lived on the edge of the city, and his experiences of Dublin were very different from those of Fitzgerald. This chapter considers the scientists' ether work as central to nineteenth-century discourse on the origins of matter, and maps the subsequent cartographies of physics that emerge from both Fitzgerald's Dublin and Barrett's Dublin

In chapter 2, I attend to the microgeography of the city, and travel to some of its institutional sites and public gardens to investigate some spaces of talk and text. This inquiry places David Moore (1808–79), curator of the Botanic Gardens in Dublin, and Alexander Macalister (1844–1919), a professor of anatomy at Trinity College, within the origins discourse surrounding the major debate on structuralism and functionalism in the life sciences during the late nineteenth century. In this debate around form and function in biology, the different requirements and expectations of their posts enabled David Moore to express his utilitarian ideology and functionalist leanings freely within the textual and rhetorical spaces available to him, and Alexander Macalister's structuralist agenda was constrained by his particular locus in Trinity's anatomy department.

Operating at yet another spatial level, chapter 3 locates Dublin within the transnational discourse on human origins and development and explores the local negotiation of this racial science through the differently fashioned habituses of the English zoologist Alfred Cort Haddon (1855–1940) and the Scottish professor of anatomy, Daniel John Cunningham (1850–1909), who together founded the Dublin Anthropometric Laboratory in 1891. This chapter documents the way congruent ideas conceived as far apart as Washington, Paris, and London came together to constitute the scientific practices and instrumentation of a universal anthropometric venture, and how Dublin's anatomical and ethnological circles posed very different questions of this new science.

From these expansive transatlantic connections we journey to the more intimate space of science's emotional geography in chapter 4 to explore language origins. This finds expression in the charged consociation between science and literature during the Irish-language revival. This chapter focuses on the lives of Romantic revivalists, George Sigerson (1836–1925) and Eoin MacNeill (1867–1945), and their respective liaisons with cultural institutions that differently highlighted the potential of science to boost national pride. It explores the ways in which emotionally heightened spaces may usefully illustrate that social relations and scientific engagement are mediated by feelings and sensibility.[67]

In this book, the city functions both as an external physical space—in terms of its institutions, streets, gardens, and proximities—and as an internal cognitive space. The word "Dublin" long had a meaning that went beyond the physical bounding of the city. It could conjure up, for instance, a sense of being on the right side of the law, or "within the pale." This sense of Dublin was vital to maintaining a separate Anglo-Irish polity or culture. On the other hand, the *Dublin Review,* a popular Catholic monthly was published in London primarily for English readers, but the choice of its name was a calculated one. Associating the journal with Dublin, identified it in the minds of the editors, with a *genuine* Catholicism, because Dublin as they saw it, was *the* center for Catholic culture. But for scientific lives, it also had very specific individual meaning. We are reminded by continuing research in the historical geography of science that "local encounters with science were shaped by how the practitioners conceived of their local settings, by what they took their situation to be, as much as by the ontological realities of those venues."[68]

The case studies explored in this book reveal that "Dublin" meant many different things to the scientists under scrutiny. For example, it emerged as Fitzgerald's "Unionist stronghold" in contrast to Barrett's "urban community in need of salvation." That same society became a candidate for a cultural renaissance in McNeill's—and to a certain extent in Macalister's—scheme of things, whereas for both Cunningham and Sigerson, the city hospitals and prisons—testament as they were to a degenerating populace—reflected differing perceptions of "Dublin" in wider evolutionary debates.[69] David Moore perceived the same city as a *welcoming* adopted home, where, for forty years, he presided with devotion over its "beautiful gardens which owe so much to his labours."[70] The scientists' different "perceptions" of Dublin therefore had a bearing on the way "origins" was differently understood and mobilized in the city. The geographical approach employed in the following chapters now reveals how the city afforded many different perspectives on the ways in which science moved within its interpretive communities.

Chapter 1

Ether

In Kipling's short story "Wireless," a young consumptive man is given a drink of chloric ether in a chemist's shop in which an experiment in radio transmission is to take place. The effects of the chemical ether and the waves being transmitted through the electromagnetic ether appear to combine to induce him to entranced automatic writing in which he seems to channel the words of the young Keats.[1]

The point of Kipling's story, as Steven Connor tells us, is to demonstrate the possibility of interference between the different registers of the ether: chemical, electromagnetic, poetic, and spiritual.[2] In the late nineteenth century, the cultural space designated as "the ether" was itself traversed, as Connor reminds us, by numerous undulations and wavelengths. And it is this tremulous territory that I wish to dwell on here. Just as ether was seen by nineteenth-century scientists to pervade all space, so too were notions of the ether pervasive in Victorian culture. This opening chapter tracks something of the local oscillations of these discussions in which ideas around the origins of matter were framed. The life-spaces investigated here concern William Fletcher Barrett (1844–1925), a professor of experimental physics at the Royal College of Science, Dublin, and George Francis Fitzgerald (1851–1901), a professor of natural and experimental philosophy at Trinity College, Dublin.

From Aristotle's ponderings on the fifth element or "quintessence" to Oliver Lodge's Victorian whirling ether machine, ideas of the ether lie behind, and make possible, some of the thinking about space that has become an integral part not just of science, but of geography and cultural studies too.[3] As Albert Einstein put it in 1929, "The ether was invented, penetrating everything, filling the whole of space and was admitted as a new kind of matter. Thus it was overlooked that by this procedure space itself had been brought to life."[4] Nineteenth-century configurations of the ether gradually became more amenable to the idea that space might give up its passive role as a mere stage for physical events.

Conceiving ether as the matter of space, or space as a kind of matter, made the notion of space mutable and productive. Thus, the ether began to be endowed with a multitude of chemical, physical, physiological, and theological functions, all of which have been amply illustrated by Geoffrey Cantor and Michael J. S. Hodge's 1981 outline of the ethers' checkered history.[5] The wide conceptual scope of their project draws initially from metaphysical and philosophical concerns, and traces the idea of the ether as an active imponderable fluid from its Newtonian origins through to the second half of the eighteenth century. The role of ether in chemistry and physiology, conceived of not as fixed form or clear outline, but as a tremulous cloud of sensations and reflexes is discussed in detail, as is its theological significance. Other studies concentrate on accounts of the mathematical and physical models of ether developed in the nineteenth century, and the scientific problems that gave rise to them, thus building on the work of the ether historian Edmund Whittaker whose foundational study on the subject was published in 1951.[6]

As one commentator has pointed out, "The editors wisely did not fall into the trap of trying to impose a superficial unity on such diversity."[7] Instead, as Cantor and Hodge note in the introductory essay, their concern was "to use the history of this unruly family of concepts to raise and clarify general issues that any comprehensive analysis of those two centuries would have to recognise."[8] This serves to alert the reader to the broad nature of the many contexts and usages that drifted into thinking about the ether when considered as a wider cultural practice of thought, and it is this domain that I wish to enter here by asking how the location of a particular physicist mattered in terms of the way ether was interpreted.

In spite of a few dissenting voices, the ether was generally regarded as indispensable throughout much of late nineteenth-century physics. Questions were raised not around its existence, but around the nature of the substance and its interaction with matter, as well as the larger metaphysical concerns that surrounded it. The issue at stake, according to the historian of physics, Helge Kragh, was whether the ether was seen as the fundamental substratum out of which matter was built, or whether matter was a more fundamental ontological category of which the ether was just a special instance.[9] Whatever the belief, Iwan Morus explains that the ether contained the mechanism through which the grand doctrine of the conservation of energy operated and became manifest. As a result, it became a compelling and authoritative explanatory tool that pervaded the scientific culture of the day. Although the ether was one of the major success stories of nineteenth-century science, within a couple of years of the century's end, the all-pervading substance was vanquished to the sidelines of physics, having been "revealed as a baroque fantasy, better fitted for the condescension of a new generation, than for any serious consideration on their part."[10] Although this chapter will intrude on the diverse terrains uncovered by Cantor and Hodge's book, the story here focuses specifically on the Dublin engagement with the ether episode during its apotheosis. It investigates the ways in which the differing locations of two city scientists in regard to ether research—whether inside or outside "establishment" parameters—influenced scientific debate.

This ether inquiry can be traced back to the Newtonian physics first taught at Trinity College Dublin in the early eighteenth century by Dr. Richard Helsham, Dean Swift's medical adviser and fellow of Trinity. Here, the course of lectures on natural philosophy presented by Helsham pondered on the problem of the stationary ether and the lack of any physical effect of the earth's motion through it.[11] Newton's dictate, in accounting for the planetary orbits in terms of laws of gravitation, meant that the very existence of the ether would disturb and retard the motion of the planets and comets. This paradox had long played on the minds of natural philosophers.

Over the next hundred years or so Trinity scholars continued to address the issue. James McCullagh (1809–47), a professor of natural and experimental philosophy, developed mathematical models for the ether but claimed that he had not succeeded in acquiring any definite mechanical conception, stating "one thing only I am persuaded of, that the constitution of the ether, if it ever be discovered, will be found to be quite different from anything that we are in the habit of conceiving, though at the same time very simple and very beautiful."[12] To comply with theoretical specifications, the mechanical qualities of the ether had to be fluid so as to fill space, but rigid enough to support the high frequency of light waves. It also had to be massless and without viscosity, otherwise it would visibly affect the orbits of planets. In addition, it had to be completely transparent, nondispersive, incompressible, and continuous at a very small scale. The Dublin mathematician Thomas Preston summed up the difficulties in 1895, when he claimed that the existence of the ether could be established only by the "intellect" rather than by direct sensory experience. Its connection with ordinary matter was "far from being settled by experiment" and there were "difficulties . . . in forming a consistent idea of its constitution and function."[13] Despite these problems, several attempts were made to draw analogies with unusual materials. Osborne Reynolds believed the ether might have had granular properties like sand. Gabriel Stokes preferred the idea of a jelly or wax. One of William Thomson's models conceived the ether as a special kind of liquid foam.[14] By the late nineteenth century, Trinity scientists along with colleagues elsewhere had made such progress in describing elastic solids and fluids that they felt ready to construct a full theory of the ether, and the ensuing deliberations were intertwined with developments in the theory of heat and an understanding of light waves.[15]

What is more pertinent to the direction of this chapter, however, is the fact that in acknowledging the existence of the ether, physicists also had to acknowledge that it was unlike ordinary matter and would have to possess extraordinary physical properties such as quasi-immateriality, universality, continuity, and unity. This made it a particularly flexible resource, which, once given scientific sustenance, could also then become a useful tool for physicists who engaged in political or religious discourse. Richard Noakes has recently made the argument that it was these supposed properties of the hypothetical ether that made it a plausible argument against a determinist and materialist cosmology, a way of comprehending Divine intelligence and Providence, a mediator between terrestrial and spiritual existence as well as a metaphor of Tory views of the British Empire, socialist views of wealth, and the spirit of international cooperation.[16]

Noakes has deconstructed Brian Wynne's much-criticized vision of a coherent "Cambridge school" view of the ether, which embodied a substantial "Strong Program" thesis.[17] He finds fault with Wynne's implied cohesiveness of the scientific group, by highlighting instead the many unacknowledged diversities that existed regarding nationality, locality, religion, and social positioning of the physicists in question. Noakes thus guides ether historiography farther into the realms of geography, and it is here that I will continue the discussion by elaborating on the role of scientific *lives* in locating ether interpretations. William Fletcher Barrett and George Francis Fitzgerald had a significant influence on the progress of nineteenth-century physics.[18] Barrett was lauded for his work on the electrical, magnetic, and thermal properties of metals and for his studies on sensitive flames. Fitzgerald played a crucial role in the development of James Clerk Maxwell's theory of electromagnetism, and for instituting the Fitzgerald–Lorentz contraction, which later became an essential part of Einstein's theory of special relativity. Barrett, a non-Dubliner in Dublin from a nonconformist background, was based at the newly established Royal College of Science. He lived on the edge of the city. Fitzgerald was a professor at Trinity, a Dublin-born Anglican, and member of an elite circle of Dublin scientists centered around the university. His experiences of Dublin were very different from those of Barrett. To better understand the complex relationship between these physicists and the cityscapes that circumscribed them, and to convey something of how their *lived* experiences in the city then directed them toward particular but different mobilizations of ether theories, the discussion now turns to investigate some of the Dublin spaces that informed the habitus of these scientists.

The Supernal Cityscapes of William Fletcher Barrett

William Barrett's view of the ether saw it more as a mode of sensitivity or susceptibility, a quasi-vital substance rather than a form of matter. Construed in this way, the ether could be deployed as a fundamental part of an argument for spiritualism. Physicists had radically diverging ideas of the ways in which religion mattered to the profile of physics in Victorian culture, and William Barrett was one of a number of new scientists who championed publicly the compatibility of physics and religion. A brief overview of that terrain here will serve to place his views in the context of contemporary religious conceptions of the ether.

For some religious physicists, the ether was useful for exposing the hollowness of materialism. Into this category would be placed those seeking to reaffirm their Christian faith and those looking for an alternative to Christianity, which they felt had been undermined by scientific naturalists. This group combined orthodox Christians, liberal Christians searching for scientific proof of religious experiences, and spiritualists, all of whom tended to be religiously motivated in their mobilization of the ether. This "persona" mobilized the ether primarily as a response to scientific materialism or naturalism, which many Victorian scientists with traditional Christian beliefs felt was a threat to their beliefs. Their fears were fueled by the publication of pure materialist dogma such as Ludwig Buchner's *Force and Matter* in 1855, and later by the implications of Darwin's compelling evidence in 1859 that all living things had developed from a few basic prototypes. The interpretations and responses to John Tyndall's controversial pronouncements of 1874 in his "Belfast Address," and the activities of the "X" club, were also pivotal in strengthening this view.

Scientists who upheld these beliefs found support in theories such as William Thomson's 1867 theory of oscillating vortex rings, the most important of a number of ideas proposing that matter itself might be a kind of convolution in the ether.[19] The ether thereby underwent an ontological promotion of sorts. No longer merely that which lay between things, it became, in the minds of some scholars, a primary matter out of which all things emerged. Following on from this, the Scottish physicists Balfour Stewart and Peter Guthrie Tait made the ether a central part of a religious argument in their 1875 book *The*

Unseen Universe, or Physical Speculations on a Future State: "We therefore welcome an hypothesis like that of Sir William Thompson, which regards the primordial atoms of the visible universe as vortices somehow produced in a pre-existing perfect fluid. . . . In the production of the atom from a perfect fluid we are driven at once to the unconditioned—The Great First Cause."[20]

Knotted vortices, according to Thomson's theory, constituted atoms. As a consequence of the work of Hermann von Helmholtz, once stamped in the "ether" at the moment of creation they would be eternal. One of the advantages of this view for the more religiously orthodox scientists like Barrett was that, although the stability of vortex rings in a frictionless medium would mean they would last eternally as atoms seemed to, they could not be regarded, for that very reason, as able to arise through some accident or evolution in the state of the ether; they required an act of precedent creation.[21]

Tait, a lifelong friend of Maxwell and collaborator of Sir William Thomson, held strong religious beliefs and with help from Balfour Stewart presented his case in *The Unseen Universe.* Their primary argument held that religion and the immortality of the soul were compatible with modern science: crucially, for scientists such as Barrett, *Unseen Universe* postulated parallel avenues of universes and described a world in which miracles and life after death were *scientifically* possible. A number of eminent scientists championed Tait and Stewart's alliance of physics and religion, but in quite different ways. For instance, Thomson's physical arguments against uniformitarianism had the theological consequence of defending the design argument against Darwinism, whereas Stokes directly employed both the design argument and biblical evidence—as reinforced by considerations of physics—to help deny the occurrence of evolution itself. [22]

Another slightly different approach was undertaken by the prominent Maxwellian, Oliver Lodge, who paved the way for Barrett. He sought to highlight the potential of physics to vanquish its materialistic image by focusing on how different ether was from ordinary matter. Lodge noted that there was a good chance that life and mind, hitherto excluded from the dynamics of matter in motion, could be accommodated "within a more general scheme of physical science."[23] Like Stokes, Lodge promulgated a universe permeated by life and mind through interpretations of ether physics and the notion of evolution "directed" by Divine agency, but he took a very different route from the traditionalist Stokes by building his mission on radical reinterpretations of Christian doctrines and appealing to the controversial results of psychical research. As Peter Bowler points out, the proponents of this view emerged as outspoken liberal Christians who challenged those aspects of Christianity that were deemed to be no longer compatible with the scientific worldview.[24]

FIGURE 2. William Fletcher Barrett (1844–1925), professor of physics at the Royal College of Science for Ireland. Pencil portrait. Reproduced by kind permission of UCD Archives UCDA RCSI/215.

For them, it meant the liberalization of the Christian message rather than a rejection of it. As one of the main proponents of this new liberal Christianity, William Fletcher Barrett was an important figure in social and scientific circles in late Victorian Dublin.

He was a moderate home ruler, philanthropist, an advocate for women's education in science and medicine and a supporter of reforms in technical education. He corresponded regularly with Oliver Lodge and Balfour Stewart while in Dublin, and one of his greatest friends and colleagues was Alfred Russel Wallace, who stayed with him on occasion in Kingstown.[25] Barrett was a founding member of the London Society for Psychical Research (SPR) and while employed as a professor of experimental physics at the Royal College of Science for Ireland, he stoked a burgeoning interest in psychical research among prominent members of the social and cultural elite in his adopted city, with the foundation in 1908 of the Dublin branch of the SPR. Despite his scientific and social standing, however, Barrett inhabited a peripheral location with respect to the conventional scientific community in 1870s Dublin. There exists no biography of this scientist, and unusually, no Royal Society obituary notice, and no biographical memoirs exist for him as a deceased Fellow.[26] He is primarily known today only in connection with the Society for Psychical Research.

Barrett's location on the fringes of acceptable science was made very clear at the British Association meeting in Glasgow in 1876, when Wallace invited him to speak to the Anthropological section. Barrett's paper on some phenomena associated with abnormal conditions of mind fueled ongoing debates, and some biologists, in particular Ray Lankester, attacked him for bringing the British Association into disrepute by "giving credence to a residuum of spiritual manifestations."[27] The peripheral publishing location of his work also weakened his reputation in the most important late Victorian scientific circles. Back in Dublin, his pro–Home Rule views—which asserted that "the solution to sectarian conflict was for all parties to experience the discipline of self-government" and be "forced to collaborate in a local legislature"—were setting him apart from the wider Unionist community."[28] This impartial stance was likely to have been inculcated in his domestic spaces where he enjoyed the nonsectarian politics of a suburban township where Catholics and Protestants often shared a middle-class social identity.[29] But Barrett despaired for the inner city, which he felt was enfeebled by the sectarian nature of politics, giving rise to situations that would be "incredible in England."[30] He found it to be inhabited by "dogmatic Protestants and Catholics who condemned all communion with spirits."[31] His nonconformist tendencies spilled over into the Society for Psychical Research as his psychical conclusions placed him increasingly at a distance from the elite of the SPR. While Barrett felt physicists were supremely qualified to investigate spiritualism, SPR members such as Henry Sidgewick and Frederick Myers were not convinced. Spiritualism with its history of associations with fraud was seen as the most dangerous area for scientific inquiry.

Some authors attribute his marginal influence on the history of science to the fact that he lacked many of the technical and social resources needed for making a scientific career and had to build his reputation and social prominence through a series of teaching posts in new science colleges, through private research, and by writing articles for the periodical press. Although this certainly describes his career path, it was one followed by many Victorian physicists such as Tyndall and Lodge whose scientific lives have been well scrutinized. The following perusal of Barrett's life path and habitus will illuminate something of his "feeling for the game" of natural philosophy, and demonstrate the ways in which he was disposed to particular actions by the promptings of particular Dublin places.[32]

Barrett was born in Kingston, Jamaica, where his father, a Congregationalist minister and member of the London Missionary Society, ran a station for saving the souls of emancipated African slaves. As Alan Gauld tells us, Barrett's father, William Garland Barrett, seems to have been successful in instilling the virtues of Christian life in his children because two of his sons also became Congregational ministers and William himself, unlike many scientific colleagues, never experienced a crisis of faith, remaining a devout and earnest Christian all his life.[33] The family returned to England in 1848 and William attended Old Trafford Grammar School in Manchester before studying physics and chemistry at the Royal College of Chemistry, London. From 1863 until 1867 he worked at the Royal Institution, London, as an assistant to John Tyndall and came under the guidance of Thomas Huxley and Michael Faraday. A frequent visitor to the laboratory where he worked with Tyndall was the Royal Institution member and Irish physician, John Wilson, father of the astronomer Edward Wilson, who invited Barrett to spend summers at the family estate at Daramona in County Westmeath. Barrett describes his astonishment in discovering Wilson to be an investigator of animal mesmerism and describes some early attempts with the Wilsons to scientifically examine aspects of mesmerism on "a sensitive subject from the estate."[34] The results of these experiments had a profound effect on Barrett and heralded the beginning of his preoccupation with telepathy.

Interests such as these were not encouraged by his mentors back in London. The majority of Royal Institution scientists had by this time developed entrenched antispiritualist attitudes and sought to exclude what was considered illegitimate and deviant from scientific investigation. Previously, an article in *The Reader* in 1864 by Tyndall had launched a scathing attack on spiritualism, deriding all associated practices, and in a series of lectures and publications in the years that followed exploited the similarity of physics and spirituality to discredit the latter.[35]

This was not, however, the only area of contention between Barrett and Tyndall. In later years Oliver Lodge reminisced that "Barrett had been associated [with Tyndall] as an assistant in his early days, where he made the discovery of sensitive flames, though there was some subsequent soreness between Barrett and Tyndall about this discovery."[36] Mollan has unraveled something of the souring of the relationship between the two physicists during 1865 and 1866, a considerable part of which revolved around the publication of experimental results on sensitive flames.[37] After some further disagreements with Tyndall, Barrett eventually left the Royal Institution to teach at a number of other London Colleges, including the International College and the Royal School of Naval Architecture. By 1867 the fledgling interest in sensitive flames developed by Barrett earlier at the Royal Institution had blossomed into substantive research on the phenomenon, and in 1868 he journeyed to Dublin to deliver a lecture to the Royal Dublin Society.[38] Here, the *Irish Times* reported that Barrett captivated his audience with descriptions of a universe that was "ringing with noiseless music," and proceeded to conduct experiments and demonstrations of this effect for the gathered crowd. He went on to talk about "complex bodies capable of being thrown into an abnormal state," which were then "sensitive to the slightest stimuli if of the proper kind." This, the report claimed was what Barrett believed to be "the foundation for whatever truth there might be in the startling facts of mesmerism."[39]

Robert Ball, the professor of applied mathematics and mechanism at the newly established Royal College of Science in Dublin writing in later years remembered the "arresting lecture which he [Barrett] gave on this subject at the Royal Dublin Society" where he "dazzled his audience by making the 'wonderful' flame bob up and down in exact synchrony to the ticking of a distant watch."[40] It was largely through Ball's curiosity about the visiting lecturer's innovative research, that Barrett obtained his next appointment. In October 1873 Ball urged Barrett to apply for the chair of physics at the Royal College of Science for Ireland (RCSI). The RCSI in Dublin, on the recommendations of a commission headed by Lord Rosse had only recently opened its doors to students with the claim that "the object of the college should be to supply as far as practicable a complete course of instruction in science applicable to the Industrial Arts especially those which may be classed broadly under mining engineering and manufactures and to aid in the instruction of teachers for the local school of science."[41]

Tyndall and Huxley were particularly zealous in placing scientific men they regarded as allies in positions of influence at the new college to help spread the teachings of scientific reform and evolution. W. T. Thistleton Dyer, a former protégé of Huxley's, became a professor of botany at the Royal College in the early 1870s and Alfred Cort Haddon was to be recommended by him for the chair in zoology at the end of the decade. Likewise Tyndall, despite recent wrangles with Barrett, supported his former assistant's move to the Royal College in a spirit of reforming zeal that aimed to establish the new scientific culture firmly in Dublin.

As these maneuverings indicate, the foundation of this Technical College in Dublin came at a time of intense debate about the teaching of science in Ireland. Whereas for Huxley, Tyndall, and their supporters in Dublin, Darwin's theory of evolution by natural selection outlined nearly a decade earlier had provided ample justification for the secularization of scientific education in the city, it was a different matter for the Irish Catholic hierarchy who viewed Darwinism as a threat to public morality. Calls to reform university education in a way that was sensitive to Catholics rang alarm bells in the minds of Darwinists, and scientific naturalists used their influence wherever possible to limit the influence of the Catholic bishops on scientific education.[42] Such was the evolutionary atmosphere that permeated the college building in St. Stephen's Green when Barrett arrived in 1873.

Darwin's theory had generated fault lines through Ireland's scientific community that initiated moves to define science once and for all. The varied doctrinal receptions of scientific naturalism in Ireland have been explored in detail elsewhere, but one pivotal event provides focus for all scholarship in the area, and this was John Tyndall's presidential address to the British Association for the Advancement of Science, in Belfast, which became an important nucleus around which science, religion, and education debates spun in Ireland after 1874. Barrett was just a year into his new post in Dublin when Tyndall's declarations were made.[43] Although the "Belfast Address" was uncompromising in its defense of the independence of the scientific enterprise, the specific Irish context for the 1874 pronouncements, and the careful qualifications of the kind of materialism he was promoting were lost in the furore that followed. In a later effort to clarify, Tyndall rejected the label of pure materialist in his own addenda to the published address, but was nevertheless interpreted as having promoted a pure materialism.[44] The reverberations of Tyndall's address were felt all over the island not just because of its inherently provocative aspects—the issues of Catholic restrictions on education, evolutionary biology, the promotion of materialism and determinism—but also because of misinterpretations of Tyndall's motives. The Catholic clergy complained that Tyndall's speech "had been carried by the periodical press into every town and village where there is a reading room or railway stall." Tyndall and Huxley were infiltrating Irish minds in the guise of, as Roger Luckhurst describes, "the Godless heads of a new intellectual order."[45] This association with materialism distanced religiously minded groups from even the outer spheres of certain scientific arenas. As David Livingstone demonstrates, the tendency thereafter in Belfast Calvinist circles to define Darwinism in atheistic naturalistic terms denied compromises such as theistic evolution in certain quarters.[46] Similarly, Mark Wood has recently shown that the Catholic hierarchy held firmly to their belief in Tyndall's "materialism" as a tool to secure a particular kind of scientific education for Catholic students.[47] That, and their failure to secure a working system modeled on their own doctrines, for many years, resulted in an educational vacuum for Catholic students.

Barrett, therefore, found himself in a city, where on the one hand, technical education—which would be economically beneficial—was not available because of local politics. And on the other hand , the university system filled the city with "young barristers, clerks and others of that class" because it was "deemed degrading to enter anything that smacks of trade or handiwork" and "great sacrifices [were] made to put children to college where they will get what is called a profession." These were just some of the socioreligious cityscapes in Dublin that convinced Barrett that promoting science in Ireland was "most important to the country."[48]

He soon realized his mission as a technical educator by mobilizing all the resources of the Royal College of Science for Ireland to superimpose Tyndall's experimental culture of empire and industry onto an industrially underdeveloped Dublin. The lure of this new culture of experimental science manifest in a progressive new college, further sweetened by Robert Ball's promises that "the opportunities for original work seem to be very great," was enough to spark Barrett's sense of possibility and grasp this chance to popularize physics.[49] For him, the act of imparting useful technical skills was a wholesome pursuit, as it also served to inculcate moral values. Barrett's religious upbringing in the nonconformist tradition with its belief that virtuous acts on earth would be rewarded in a future life had shaped the nature of his teaching and anchored his habitus in such a way that carrying out civic duties in Dublin "went without saying."

Over the next few years Barrett was responsible for launching Dublin's first systematic classes in practical physics, constructing the Royal College of Science's first physical laboratories, effecting an overall growth in staff and students, and promoting physics to a subject in which students could graduate. The college gained a reputation for the excellent equipment of its laboratories and the emphasis on the practical teaching of science was deemed to be an inspiration to the rest of the country. Barrett's quick assimilation of contemporary scientific developments also played a major role in the broadening of the college's influence beyond simple teaching to the world of commerce, industry, and medicine.[50]

As well as his commitment to college duties, Barrett found time to carry out extensive experiments in his primary area of research: sound and vibration. During the 1870s in Dublin, he was commanding large audiences for his lectures on topics such as sensitive flames and glaciers, and in the public arena he was avidly promoting Huxley's image of the new science and admirably filling the role of Tyndall's protégé. In 1878 it was reported that "the science worship which is the religion of the hour, reached the pinnacle of its popularity last night in the conversazione" following Barrett's display of sensitive flames for delegates of the British Association meeting in Dublin.[51]

However, Barrett was also simultaneously carving out a niche that distanced him intellectually from his old teacher and his Royal Institution colleagues. As well as his formative experiences with mesmerism, Barrett's religious background had also provided him with the theistic space to allow dalliances with the spiritual; in his words, "Science reveals the garment of God, religion the heart of God . . . they are one in origin, and therefore in the progress of science we ought to see more clearly the existence of spiritual laws in the natural world."[52]

His continuing interest in matters spiritual led eventually to the setting up of the London Society for Psychical Research in 1882, the American Society for Psychical Research in 1885, and the Dublin Branch in 1908. The aim of these organizations was to make a systematic attempt to investigate debatable phenomena designated by terms such as "mesmeric," "psychical," and "spiritualistic" that amid "much allusion and deception" were "prima facie inexplicable on any generally recognised hypothesis and which if incontestably established would be of the highest possible value." Barrett contributed most to those areas of research that depended more on skills in experimenting on and isolating real-time effects, the kinds of effects encountered in physics laboratories. In Dublin, psychical research attracted mostly members of the social and cultural elite. In 1875 it was claimed that there were "more than twenty clergy men in Dublin, to say nothing of certain scientific and philosophical professors in connection with the leading scholastic establishments of Ireland, who are not only willing but anxious to investigate spiritualism."[53]

The Dublin branch of the SPR was an ethereal space of fashionable society, local intelligentsia, and Protestant free thinkers, described by one member as "a research group of intelligent, informed and highly placed men and women."[54] The organization worked hard to create a "respectable" scientific image and maintain that they were not "spiritualist" societies per se. Barrett's activities occasionally led to disagreements with members. His interests encompassed some of the many areas of what was castigated as pseudoscience and supernaturalism, being sidelined by the new generation of scientific naturalists who, as we have seen, derided and discredited spiritualism and all associated practices.[55]

In 1875, two years into his Dublin post at the Royal College of Science in St. Stephen's Green, Barrett embarked on a review of three recently published books by well-known investigators of spiritualism, William Crookes, Alfred Russel Wallace and Asa Mahan, which was to appear in the *Nonconformist*, a little-known periodical of the time. The book review proved so popular that he was compelled to write an expanded version three weeks later. Although this was ostensibly a critical review, it served too as the platform from which Barrett launched his own views on spiritualism and specifically targeted what he also saw to be Tyndall's materialistic message the year before, and it is this latter aspect that interests me here. Because it was here that he mobilized the ether in order to bring together his disparate areas of physical and psychical research, and thus to mount an attack on materialism.

Barrett viewed the ether as a vehicle by which the universe could once more be seen as a *unified whole* following the advances of scientific naturalism. His was a "persona" disposed toward alleviating the woes of those "yearning for some deliverance from the meshes of materialism" and "groaning beneath a mechanical universe." Barrett, like Oliver Lodge, sought to bring the psychical and spiritual within the realms of physical science by underlining how different ether was from ordinary matter. As he explained, the ether illustrated the "transcendant unity of nature" whose true significance lay in the unifying Divine mind underlying such "material" links, and the ether seemed to have none of the imperfections associated with matter but had attributes of the Divine—that is, a perfect continuity and capacity to be the "universal connecting link" of the cosmos.[56] Because of his conviction that by using the latest techniques of experimental physics one could prove the independence of mind and body, the ether became particularly useful to Barrett in emphasizing the necessary connections between his physical and psychical research.

Although most controversies in the area of science and spiritualism were ostensibly disagreements over the reality of psychic phenomena, they were also conflicts over what kind of expertise was considered appropriate for investigating phenomena that were simultaneously physical and spiritual. On the face of it, Barrett's experimental methods and devotion to physics fitted well with the new order in science espoused by Tyndall and Huxley, but the psychical and spiritual meanings conveyed were "beyond the pale" as far as the "X" club savants were concerned. Ironically, it was Tyndall's experimental culture at the Royal Institution that played a major part in Barrett's construction of a science of mesmeric and spiritualistic phenomena.[57] Tyndall had trained Barrett on the use of numerous instruments and resources for manifesting, manipulating, and measuring vibrations in the invisible ethereal medium that were often beyond the range of human vision. His work with Tyndall had developed an acute appreciation of synchrony, and in an issue of the *Quarterly Journal of Science* he expanded on analogies that Tyndall had drawn between light and sound, and between human organs of sense and musical instruments. This was a first step toward his later claim that human perception of light and sound was the result of "sympathetic vibration" between the acoustical and luminous waves and the fibers of the inner ear and the rods and cones of the eye. By the time he was ensconced in Dublin, Barrett made this branch of physics fulfill religious and spiritual functions and embarked on a quest to take Tyndall's methods in a new direction in his adopted city despite opposition from some quarters.[58]

While he was employed at the Royal College of Science, Barrett actively pursued the idea that telepathy might be comparable to physical systems exhibiting resonance. By the 1890s his article "Sympathetic Vibrations" in the Christian publication *Good Words* started with a survey of resonant phenomena in acoustics and concluded with the observation that sympathetic vibration was a principle that held for the invisible vibrations of the ether and supported "many obvious spiritual analogies."[59]

Barrett held that mesmeric and spiritualistic phenomena were puzzles that physicists could interpret and investigate more effectively than psychologists or physiologists. Because he felt that materialism posed a bigger threat to faith than spiritualism, it was important to him to apply proper investigation that could be used to discredit materialism and he fought hard to make psychical research a branch of physics. In his article for the *Nonconformist* Barrett explained in great detail that by using the latest techniques of experimental physics one could prove the independence of mind and body. This was of utmost importance, because he sensed that "in the hands of the superstitious, simply curious and ignorant, spiritualism was the path to moral derangement and impiety," but when it was conducted "in the spirit of an honest search for truth" and considered by "the dry light of science," it provided objective proof of "facts of transcendent importance," and a source for "stirring the potent conviction that there *is* a spiritual body."[60] In his RCSI lectures he espoused a connection between the physical and spiritual worlds and emphasized that research on phenomena manifest to the senses, whether glaciers, sensitive flames, or "spirit writing," showed the existence of spiritual laws in the natural world.

This tangential position was further underlined by the siting of his home base in Kingstown, where the physical distance from the coterie of scientists in the city left him beyond the bounds of the "cognitive topography" of Dublin physics.[61] Added to this, Barrett's psychical interests were further ignited when James Wilson, the brother of his old colleague from Westmeath, became a neighbor in the coastal suburb. Throughout the 1870s he attended séances held in Wilson's home. Barrett also engaged in experiments here with the Lauders family, who under the name Lafayette, were spiritualists and leading photographers in late Victorian Dublin.[62] On his summer vacations, he made extensive investigations of dowsing in neighboring County Wicklow and published two book-length articles on water divining in the proceedings of the Society for Psychical Research in 1897 and 1900.[63]

We can now begin to see something of how Barrett's Dublin spaces actively encouraged and endorsed his particular understanding of the ether. Returning to the crucial tenet of his faith—that religion and the immortality of the soul were compatible with modern science—it becomes clear that he was strongly enabled by his Dublin locations to further develop the connections established in London between a little-known aspect of natural philosophy and an uncommon approach to the spiritual to underpin such a thesis. His research area of choice in physics—the study of "sympathetic vibrations"—was an obscure part of Victorian physical science, the results of which lacked consensus in the scientific community. Similarly, his reasoned "scientific" approach to the study of spirituality was unconventional. Located in the newly established Royal College of Science in Dublin, removed from the watchful eye of Tyndall and distanced from the natural philosophers at Trinity College, Barrett's borderline scientific endeavors flourished and his pedagogical mission to promote the spiritual and moral value of physics was uncurbed.

Barrett's dedication to teaching both day and evening students in Dublin reflected his strong belief that physics was the only way to revitalize faith in Christian spirituality and safeguard public morality. His educational role outside the Royal College of Science whether in lecture theaters, schools, or articles in popular magazines was also part of this mission to popularize the subject. To him, physics was a "means of *education*" rather than a mere "vehicle of instruction," and when properly taught it "educated individual judgement by training the senses to habits of accurate observation and the mind to clear and precise modes of thought."[64] In the 1870s his public role in Dublin, although similar to that enjoyed by Tyndall in London, projected a very different message. Barrett's popularizing zeal limited his standing in the emerging scientific societies, but it increased his power over public audiences to whom he wanted to promote the image that physics, because it revealed nature's mysteries, was a form of religion. This intellectual space was further bolstered in his attempts to build up a reputation by frequently publishing his findings in religious periodicals such as *Good Words*, the *Nonconformist*, and *Light*.

His time in Dublin therefore, which he described later in a letter to Oliver Lodge as "the most God forsaken spot on this earth," provided him with ample space to instill moral values as well as teach useful technical skills.[65] It was by showing that physics was full of mystery and wonder, not merely reductive and mechanical that he succeeded in emphasizing the role of the experimental physicist as spiritual educator. Dublin's urban poverty and social strife elicited some of his most impassioned outbursts. In despairing letters to colleagues, he spoke of the "bottomless misery and poverty of the infinite drinking classes," which he felt were caused in part by the brewers, distillers, and publicans whose commercial success won them seats in the House of Lords.[66] As "Barrett's Dublin" was a city in which urban poverty and social strife called out for salvation from the excesses of commercialism and materialism, he devoted himself to a range of worthy causes from ameliorating the conditions of the poor to suppressing religious intolerance. During his tenure at the Royal College, he also established a nonsectarian teetotal men's club, promoted free libraries, women's education, and temperance.

Viewing the city from the perspective of temperance halls, working men's clubs, and relief centers allowed Barrett to call on what he saw as the edifying qualities of physics to remedy social disharmony. The ether was vital to Barrett for bridging the physical and psychical worldview, and his understanding of the substance as "the primary instrument of the mind and the habitation of the soul" enabled him to fashion a scientific space where both physics and religion could reside. Ether physics therefore became part of his moral quest to improve the lives of Dubliners.

So we see that the spaces Barrett occupied in Dublin facilitated his passion for the intellectual, moral, and cultural importance of physics and led him to believe, unlike scientific naturalists and orthodox Christians, that spirituality played a part there. His peripheral locations in Dublin allowed a particular mobilization of the ether that helped to shape the institutional and pedagogical spaces around him, and in so doing, influenced the scientific culture of the early years of the RCSI in a manner that ran counter to the prevailing climate of science.

Mobilizing the Ether in George Francis Fitzgerald's Dublin

At the time of Barrett's publication in the *Nonconformist*, his colleague at Trinity was also contemplating the ether, but it would be another three years before a chance introduction would influence the course of his ponderings. In 1878 Dublin was the venue for the British Association meeting, and it was here that a visiting Oliver Lodge, soon to be a physics professor at the Univer-

FIGURE 3. George Francis Fitzgerald (1851–1901), professor of natural philosophy at Trinity College Dublin. Wikimedia Commons: Public domain image.

sity of Liverpool, met the young Dublin physicist George Francis Fitzgerald. The two young men found they had much in common, including a shared enthusiasm for Maxwell's theory and a conviction that all electromagnetic phenomena could ultimately be traced to the strains and motions of the ether. They began an active correspondence and exchanged regular visits across the Irish sea, often meeting with other scientists such as Joseph Larmor, Heinrich Hertz, and Oliver Heaviside, and ultimately because of their shared interests, formed a group that later became known as "the Maxwellians." As their constant source of advice, encouragement, and bold ideas, Fitzgerald assumed the role of "St. Paul" among these "apostles of James Clerk Maxwell" according to Denis Weaire and Michael Coey.[67]

As a native son, a new man of science, and ardent Unionist, Fitzgerald's "Dublin" was, first and foremost, the second city of the British Empire.[68] He was an Anglican and member of the established Church of Ireland. Central to his vision for the advancement of science was the acknowledgment of this imperial status and the primacy of Trinity's place in it. Fitzgerald's eagerness to both encourage the economic benefits of the growing electrical, chemical, and telecommunications industry and secure Dublin's role as a venue for science can be best understood in the light of recent observations by the cultural historian of science Iwan Morus. Here, we are informed that "in civilisation, in industry and in the production of knowledge, 'progress' was the buzzword of the Victorian age."[69] And increasingly, during the late nineteenth century, science was tasked to harness nature's machinery in the name of this material progress. The attempt by late Victorian Liberal governments to grant Home Rule to Ireland was anathema to such views of imperial science. The proposal was seen by many as a potentially retrograde step that would undermine technological progress. It becomes increasingly apparent as we investigate the habitus of Fitzgerald, that the social and cultural capital (in the Bourdieuan sense) that formed his intellectual response was firmly rooted in the collective of scientists, outlined by Greta Jones who opposed Home Rule. A local Dublin counterculture influenced by Catholic hostility to the "Protestant modernisation" of science and Nationalist pressure over academic appointments was set to undermine this unified scientific elite. It was feared that because Irish science was sustained by the close relationships between British and Irish scientific communities, agitation of a Nationalistic hue would disrupt these scholarly relationships, to the detriment of Ireland.[70] My intention throughout is to probe the dynamic aspects of Fitzgerald's interacting political, religious, and intellectual spaces and reveal a habitus largely cultivated by ways of thinking that prioritized conceptions of unity and continuity over those of division and disjunction.

At the same time, as we have seen, Dublin scientists were contributing to the advance of natural knowledge in another important nineteenth-century arena. Here, it was ether physics that excited the imagination of scholars and provided the framework within which future progress in science would take place. To this end, Dublin's Royal College of Science has been disclosed as a space in which the "psychical and spiritual" could be brought within the realms of physical science, aided by Barrett's appeal to the differences between ether and ordinary matter. The chapter now turns to Trinity College Dublin and to a mobilization of the ether that strove instead, to underpin Protestant orthodoxy. In contrast, the focus here is on a scientific life that pursued connections between ether and matter.[71] For a start, Fitzgerald entertained the idea that the ether was a connecting link between the sciences themselves, thus blurring the spatial boundaries between empirical science and the workings of the "scientific system." He believed that the ether demanded "elucidation and strength," because without it, scientific disciplines were in danger of becoming too specialized and would suffer from "undernourishment" from other disciplines, and "local turgescence and inflammation would damage the whole scientific system." But Fitzgerald took this notion further in a revealing account of his thoughts on the subject of science and society published five years before his death. Here, he made a striking analogy between the corporate life of animals, social states, and the organization of the sciences where he argued that science was progressing in the same way as civilized states: "In general it needed intercommunication between its different disciplines, and in particular it required that research done in physical science be collected, digested and distributed to the biologist, chemist and geologist." The study of the "properties of each kind of matter as related to energy and the ether" was the most pressing aspect of this research because it had "bearings on every department of science and on every practice."[72]

Connections such as this one made by Fitzgerald between the ether, scientific disciplines, and social states have led Richard Noakes to suggest that Fitzgerald represents one of the few "plausible cases of a late Victorian physicist implicitly teasing out the social implications of the unifying capacity of the ether." Noakes's observation—that "corporate life" might have been a veiled reference to the union of Great Britain and Ireland that was the preserve of the conservative Cambridge elite of which he was a member—is intriguing. [73] By the 1890s, we know that Fitzgerald was fiercely opposing Home Rule and was only too aware of what he considered to be an undermining of the progress of a "civilised state," that is, the attempt by Nationalists to disengage with the British Empire. So we see that in tandem with Dublin's esteemed imperial position being threatened by Nationalist aspirations to disengage with Britain, a general intellectual propensity to foreground notions of unity and cohesion emanated from Fitzgerald's scientific assertions on the ether. Whereas the substance—which was useful to Fitzgerald as a tool for supporting his belief in the unity of nature and the sciences—can be seen as a metaphor for his preferred polity as Richard Noakes outlines, my aim here is to approach from the counterpole and educe the extent to which tradition and ideas generated locally drifted into his scientific work.

Fitzgerald's preoccupation with this particular branch of physics, and his reluctance to accept the demise of the ether, I argue, was the manifestation of a specific Dublin habitus in which moves to accommodate modern science within the framework of Anglican theology were intrinsic. The quest to establish and underpin the natural links between "modern science" and Protestant culture was a vital component in strengthening the anti–Home Rule position. Fitzgerald therefore found in ether physics a crucial space in which to maintain a credible working relationship with scientific modernizers such as Tyndall and Huxley without forsaking the traditions of his Dublin Christian heritage, while at the same time reinforcing his anti–Home Rule message.

To explore these points the discussion congregates around three interconnecting but loosely sequenced themes. These introduce Fitzgerald's many Dublin lives; scrutinize his role as a scientific modernizer; and finally probe the essence of his anti–Home Rule stance. First, as a Dubliner, the *personal* stakes were far higher for Fitzgerald than other anti–Home Rule Unionist colleagues at Cambridge, such as James Prescott Joule, Lord Rayleigh, Balfour Stewart, Peter Guthrie Tait, and Joseph John Thompson. Any curtailment of the great imperial project by the establishment of an Irish Parliament, would mean the complete uprooting and expiration of his many Dublin "selves," as we shall see. Second, by investigating Fitzgerald "the physicist," as one of the new breed of late nineteenth-century "scientific modernisers," we can, by comparing his intellectual position with those of aggressively anti-Catholic "naturalist" colleagues such as Tyndall and Huxley, unearth something of the agency of his native city—particularly Trinity College—on certain tenets of natural philosophy that were fundamental to late Victorian ether theories, but were grounded in Christian theology of a Hibernian stripe.

Third, by acknowledging Iwan Morus's observation that physics at this time was the vehicle for the expression of deep fears about what the new century had to offer and Greta Jones's claim that the cause of scientific modernization and (Social) Darwinism were truly connected; it is my contention that further scrutiny of Fitzgerald's Dublin life-spaces allows for a better understanding of the depth of his feeling about the breakup of the Union in the context of contemporary fears; not just about the role of science in a post–Home

Rule Ireland, but also around the nineteenth-century concepts of progressive evolution and degeneration.[74] Reflecting along these lines brings Fitzgerald's fin-de-siècle ruminations on ether physics into sharper focus.

The elevation of physics as a scientific discipline and the widening of its authoritative jurisdiction were well under way in the 1890s. Its practitioners were viewed as rational modernizers who drove the all-important advances in technology. But the negotiating grounds on which scientific expertise and authority were debated had been appropriated by "X" club members, resulting in a conflation of "scientific modernism" and "naturalism." Ether physics, however, remained a compelling area of research for both theoretical and experimental endeavors in physics because it provided space for scientists like Fitzgerald who were not antiestablishment or antireligion to locate themselves within the modernist movement in science. For many of these more establishment and religious scientists, the notions of unification and progress were inseparable, and for them the ether can be seen as having evolved as an inevitable outcome of the human quest to understand the cosmos in a way that upheld Protestant traditions of both science and society.[75] For Anglicans such as Joseph Larmor, whose writings contained much historical analysis and many hagiographical accounts of Protestant ether builders, the ether symbolized the importance of traditional values over radical change in science and politics.[76] For Fitzgerald, however, this was far from a theoretical abstraction, He was, by contrast, *living* through political change and uncertainty. The immersion of his many Dublin personae in the changing political landscapes of Unionist Dublin, was I suggest, effectively reinforcing a traditional approach to his science. To illustrate this, it is first necessary to inspect his crucial relationship with the city.

Like Barrett, Fitzgerald believed that Irish society was in need of social reform and improvement, but his perspective was formulated through a very different lens. From a city perspective, and in sharp contrast to Barrett's peripheral status, Fitzgerald was a figure whose physical, intellectual, and social presence radiated from a central location. If we accept that scientific pronouncements are defined by reference to the positions—the moral and social spaces—from which they speak, then Fitzgerald's place in the city matters a good deal in trying to understand his mobilization of the ether.

His was a position of social and scientific authority in a city that was central to his adult life and work. Fitzgerald "the scholar" remained at Trinity from the age of sixteen until the end of his life, while others had left for posts abroad. He was immersed in a culture that was comfortable and familiar to him. The extended Fitzgerald family was embedded in the fabric of the college; his father and uncle were Trinity professors; and Fitzgerald was married to Harriet Jellet, the daughter of the provost of Trinity. Their home at No.7 Ely Place, was a short stroll away from the college.

Outside the radiant rooms of Trinity, the grounds of the university were not simply the bounds of his workplace. On the greens of college park, Fitzgerald "the aeronaut" constructed, and was hauled skyward on a Lilienthal glider by his students. Nearby, Fitzgerald the "sportsman" practiced pole-vaulting and hockey and oversaw the affairs of the boat club. The performing Fitzgerald spent his evenings with eminent Dublin families of the time, such as the Jellets, the Stokeses, and the Jolys enjoying music and amateur dramatics in Dublin's Georgian drawing rooms.[77]

In his college, Fitzgerald was regarded as the "idol of the undergraduates and the hope of the older men."[78] Here, he also founded and became the most industrious member of the Dublin University Experimental Science Association, which met once a month to discuss and present new work over tea. On one of these occasions we are offered a rare contemporary glimpse of Fitzgerald's authoritative persona. A young Robert Lloyd Praeger reflecting on his time in Dublin in the early 1890s mentions that "the interest of those years was heightened by the existence of a little coterie of scientific men who met daily in a modest bun shop in Lincoln place opposite the gates of the college." As a junior member he was welcomed into the conclave, and described the activities of the group:

> [They] drank tea and devoured vast quantities of buttered toast and the discussion was on every subject within the confines of the universe—and sometimes outside them. Our leader and Chairman *honoris causa* was George Francis Fitzgerald, and around him were grouped W. J. Sollas, A. C. Haddon, Grenville Cole, Thomas Preston, Frederick Trouten, R. J. Moss and—John Joly would drop in occasionally, Sir William Ridgeway and Sir Joseph Larmor would join us if they were in Dublin. It was always a lively lunch table, the genial Fitzgerald essentially a product of Dublin and Dublin University, was a big bearded man, and his resonant voice filled the little room as he poured out wisdom and nonsense in a delightful medley—for he was a brilliant talker—often to the astonishment of customers at other tables.[79]

Commenting on the powerful influence of the new men of science in the late nineteenth century, the words of the economist and social reformer Beatrice Webb could well have applied to Fitzgerald when she observed, "Who will deny that the men of science were the leading British intellectuals of that period; that it was they who stood out as men of genius with international reputations; that it was they who were the self-confident militants of the period, that it was they who were rousing the theologians, confounding the mystics, imposing their theories on philosophers, their inventions on capitalists, and their discoveries on medical men; while they were at the same time snubbing the artists, ignoring the poets, and even casting doubts on the capacity of the politicians."[80]

Fitzgerald epitomized this scientific persona or "creature of historical circumstance" alluded to by Webb that laid claim to cultural authority in the late Victorian era. [81] The power shift to these new men of science appeared in the 1870s with the emergence of an ideological "settlement" that grounded public assertions of science *outside* the compromises with church orthodoxy. The movement had its most forthright expression in the pronouncements of the London-based "X" club—a group of scientists who shared ideological commitments to "untrammelled science and liberal politics," and exploited connections with influential editors of the periodical press. These men became one of the most vocal and visible groups on the Victorian intellectual landscape and plotted an aggressive campaign to reclaim nature from theology and to place scientists at the head of English culture.[82] Ostensibly, Fitzgerald allied himself with this group. He joined with the new professionalizers in their attempts to promote science and technical instruction wherever possible, and like them, he upheld the moral and utilitarian benefits of such a training. He also attended meetings of the Physical Society of London, an organization that identified itself as a modernizing influence on science.[83] But the "naturalism" associated with the new movement was not embraced by Fitzgerald. In this respect he shared common ground with physicists Maxwell, Tait, and Thomson, characterized by Crosbie Smith as "the North British defenders of the faith."[84] Fitzgerald's relationship with the new science and its main proponents—such as his Irish colleague John Tyndall—was more nuanced, however, due to his particular location in Trinity, and deserves a brief outline here.

As Luckhurst outlines, scientific naturalism in colonizing the hitherto theological terrain laid claim to an increasing number of intellectual and institutional spaces as it progressed.[85] For instance, in the early 1870s Huxley became president of the Royal Society, and in 1874 Tyndall became president of the British Association for the Advancement of Science. In addition, Huxley's *Lay Sermons*, Tyndall's *Fragments of Science*, John Lubbock's *Origin of Civilisation*, and Edward Tylor's *Primitive Culture*, were foundational assertions, and along with Darwin's *Descent of Man*, these texts had a significant influence on intellectual culture of the late Victorian period. As Greta Jones observes, "The ethos of this scientific modernism was wrapped around the pill of scientific naturalism,"[86] and because in some circles this naturalism was conflated with materialism, Fitzgerald was tasked to find accommodation for his Anglican faith within the new science. However, by encroaching on previously theological terrain, the movement also opened up new space for debate within orthodox Christianity on the relationship between mind and matter. In Fitzgerald's own discipline of physics, the establishment of connecting links between ether and matter not only answered questions of science but also advanced wider societal concerns. The ether became the ultimate vindication of the new physics: "It was real because it could be manipulated—it could be made to do things."[87]

When Tyndall in 1871 reflected the scientific naturalists' view that the "foremost men of the age accept the ether not as a vague dream, but as a *real entity,* it was a substance endowed with inertia, and capable, in accordance with the established laws of motion, of imparting its thrill to other substances," he demonstrated the extent to which scientific naturalism *could* "embrace entities whose existence could not be empirically observed."[88] In fact, the new ethereal worldview was not incompatible with the preceding mechanical worldview, but it nonetheless differed from it by its emphasis on continuity rather than discreteness, and the corresponding primacy given to ether over matter.[89] In this respect, Fitzgerald and Tyndall had no argument.

In political terms, the adoption of the ether as an aid in befogging the boundaries between life and nonlife, and mind and matter, was propitious to the naturalists' aim of marginalizing the churches' role in scientific debate. To drive this message home, Tyndall's "Belfast Address" in 1874 was unambiguous in its support of a link between the law of the conservation of energy and evolutionary theory. Here both he and Fitzgerald began to diverge in their opinions. Tyndall preferred a disciplinary alignment quite different from that of Tait and Balfour, by using the ether to tether physics tightly with biology, before taking aim at the church. Because living things were subject to energy conservation the same as nonliving, the law helped Tyndall eradicate the distinction between life and nonlife. The law of the conservation of energy was, according to Tyndall, that which "binds nature fast in fate" to an "extent not hitherto recognized, exacting from every antecedent its equivalent consequent, from every consequent its equivalent antecedent, and bringing vital as well as physical phenomena under the dominion of that law of causal connection which, so far as the human understanding has yet pierced, asserts itself everywhere in nature."[90]

Although for Fitzgerald, there was no dualism between mind and matter and, like Tyndall, he envisaged the ether as a dynamic continuum, Tyndall's deterministic nebular hypothesis was a step too far for him. Fitzgerald believed that the complexity of the ether was insignificant compared with that of organic systems and held that the dynamic laws he believed completely described the ether were not arguments for biological determinism and materialism. His friend and colleague Joseph Larmor summed it up by saying that "mechanical determinateness" need not involve "molecular determinateness" because the mechanical principles that were so useful for understanding the ether and other systems undergoing no structural change could not be employed to understand the molecular changes causing the origin and development of organic systems.[91] Fitzgerald's own language appears to be more dismissive of the materialistic element in evolutionary biology; he considered certain biologists to be "self-sufficient fools" for insisting that nature including human life had to follow fixed laws.[92] This term applied also to radical scientific naturalists in his own discipline of physics, and his arguments against materialism had their genesis in reaction to Tyndall's provocative Belfast pronouncements. The negative reaction to materialism was fast gathering pace in both scientific and public spheres, as a *Record* commentator had put it, "[Tyndall] accepts that the theory of evolution requires us to imagine not only that all the structures, animal and vegetable, were once potentially present in the fire mist of the nebulous theory, but also that all mental powers [all our philosophy, all our poetry, all our science, and all our art]—Plato, Shakespeare, Newton, and Raphael—are potential in the fires of the sun."[93] Despite presenting to the public a "radical mechanistic" persona who propounded "materialistic" summations of life's origins, Tyndall did not hold the view that all references to mental states merely described material events in the brain.

Several historians have discussed the new scientists' particular preoccupations with the spiritual and transcendental aspects of nature and, in particular, Tyndall's recognition that although the laws of matter and energy were the most reliable descriptions of the physical world, they were insufficient to answer profound questions such as the origins of life, force, and matter.[94] Even though Tyndall's mechanistic leanings suggest an origin in a broader idealist metaphysic, his mobilization of the ether as part of the nebular hypothesis to illustrate a deterministic version of "atomism"—in order to undermine church authority—was anathema to Fitzgerald, whose Dublin spaces buffered him from taking such a reductionist view.

Although other scientific naturalists such as Huxley had chosen empiricism over idealism, Fitzgerald's writings do reveal a position similar to that of Tyndall in which he sets scientific materialism in the larger context of natural supernaturalism.[95] But, whereas Tyndall found inspiration in the works of Thomas Carlyle and German Romantic writers such as Fichte and Goethe, Fitzgerald's idealism emerged from the older mathematical and Divinity traditions of Berkeleyan philosophy.[96] Fitzgerald makes clear that his interpretation of Berkeley provided fundamental metaphysical support for his contention that the physical world was reducible to pure motion. Bruce Hunt has suggested that although Fitzgerald did not specifically connect the ether to metaphysical or religious questions, his attempts to reduce it to a form of pure motion in an incompressible fluid owed much to his belief that "following Bishop Berkeley, the cosmos reduced to forms of motion that were objective manifestations of a Divine thought."[97] Fitzgerald devised his vortex sponge model of the ether in 1885, inspired by Thomson's vortex mechanics and discussions with his uncle and the Trinity physicist George Johnstone Stoney.[98] Stoney had shared Berkeley's view that the world of phenomena, and the motion of the elemental ether in particular, was a manifestation of the thought of God. His belief that the "elemental ether was space itself, regarded as moveable" had helped Fitzgerald formulate his theory and underpin his contention that atoms might be stable vortex rings in a perfect liquid ether, making *matter* itself simply "a mode of motion" of the all-pervading ether.[99] Fitzgerald believed his model would be the greatest step "towards the comprehension of the intrinsic structure of the Universe which had been made since the time of Newton." It was hailed as a triumph initially by fellow Maxwellians. The success of the model was important for Fitzgerald because of his belief that there was a more fundamental level of reality beneath the laws of phenomena outlined by Newton and Maxwell.[100]

Fitzgerald referred to Berkeleyan philosophy in his Helmholtz Memorial lecture published in 1896 but had already outlined his thoughts quite succinctly in an 1890 lecture to the Royal Institution, where he made a more explicit connection with his vortex sponge theory: "This hypothesis explains the difference in nature as differences of motion. If it be true, ether, matter, gold, air, wood, brains are but different motions. Where alone we can know what motion in itself is—that is, in our own brains—we *know* nothing but thought. Can we resist the conclusion that all motion is thought?"[101] The fact that these elements of Berkeleyan philosophy permeate Fitzgerald's explanations of the ether can be attributed directly to his location at Trinity College Dublin.

The philosophies of Berkeley, himself a graduate of the university and member of the Anglo-Irish tradition, had a lasting legacy on Anglican thought emanating from Dublin's oldest university. With a name meaning "holy and undivided," and founded in a dissolved monastery, Trinity College was a major center for religious education, and the presence among its fellows of many clergy meant the established religion of the Anglican Church was never far from the mind of Fitzgerald and other late nineteenth-century Trinity practitioners. A sense of the enduring admiration for George Berkeley circulating in the college's corridors of physics emerges from a foundational piece on idealism published in 1872 by Fitzgerald's colleague, William Graham. In his essay, Graham claimed that the philosophy of George Berkeley was the only hope for the future spiritual life of humanity because Berkeley offered the most complete antidote to these new "theoretical debasers of the human spirit," and he went on to claim that "materialism debases the human spirit because it makes thought itself a function of the brain and treats the human being as nothing more than a highly elaborated physical organism." In Graham's view, Darwinism was the most pressing variety of materialism in the 1870s, as it tried to "trace an unbroken continuity from protoplasm to man, from chemical action to the energy of the soul, even filling up the gap that was assumed to exist between vegetable and animal life." He despaired because he was sure that a spiritual life worthy of the name "cannot flourish in a materialistic ethos," and emphasized that for the idealist, it is not physical matter but consciousness that is the most important thing.[102] Finally, commenting on the legacy of the philosopher, he states that "for Berkeley, conscious intelligence, either subjective or Divine was from the beginning, is, and ever must be the Universe."[103]

It is evident too that friends of Fitzgerald in the Divinity School, such as Charles Frederick D'Arcy (later Archbishop of Dublin), were led toward these idealist philosophies as a bulwark against the materialistic interpretations of Huxley and Spencer. As D'Arcy tells us, "I felt a strong desire to know why distinguished men of science could no longer believe" and "Huxley, Tyndall and Spencer were the names that excited most horror among the orthodox." In the same publication D'Arcy also reveals the influence of Tyndall's challenge to established views and underpins the extent to which Berkeley's philosophy had infiltrated the Trinity viewpoint: "As I followed his [Spencer's] efforts to exhibit the process of the universe as a great evolution, I began to discern the nature of the problem which had to be faced. Tyndall's Belfast address was, however a more direct challenge; and I remember how I turned to Lange's *History of Materialism* to get some light upon it. Then it was that there came back to me the vision of Berkeley's philosophy which I had got in my early youth, and I was able to see how to meet materialism."[104] The mathematical and physical science departments to which Fitzgerald was affiliated had close ties with the Divinity School. Charles Frederick D'Arcy was well acquainted with the physical scientists, and mentions in his autobiography that "among the young men who impressed me were especially Fred Purser and George Fitzgerald, the researches of the latter it has proved led the way towards the new Doctrine of Relativity." R. H. Murray recalls also how J. B. Bernard and George Fitzgerald were always on "friendly terms," they had "mathematical and experimental science tastes in common, and they both manifested a lifelong reverence for the philosophy of Bishop Berkeley."[105]

In the late nineteenth century as contemporaries of Fitzgerald such as the mathematician George Salmon were making bold statements by abandoning geometry for theology in what appeared to be a growing confrontation between the new science and religious belief, Fitzgerald was less conspicuous in his religious convictions. But in 1901, J. B. Bernard, in preaching the physicists funeral sermon, threw some light on the matter: "He would not divorce knowledge from life, for he knew full well the imperfection of our best attempts at theory. George Fitzgerald was a deeply religious man, though he did not speak much in public about religion. He knew as much as anyone of the secrets of nature: he did not waver in his belief that nature is ordered by one whose way is in the sea and whose path is in the great waters, though his footsteps are not known." And Bernard continued: "Do not speak to me of natural and supernatural, he said once, there is no difference really. You must not seem to banish God from nature."[106] Fitzgerald's reluctance to be drawn on matters religious reflects the problematic space he inhabited between the mores of Protestant theology, and the pantheistic ideologies of scientific modernizers like Tyndall. Unlike these "quasi-secularists" who were attempting to disengage from the church, Fitzgerald sought to strengthen connections between the Anglican establishment in Dublin and the "new science." Apprehending the situation from a spatial perspective, we can appreciate more fully how the flexibility of the ether—in advancing idealistic viewpoints of varying shades—made it an important scientific refuge for Fitzgerald. As a consequence, his vision of Dublin as a technological Utopia, generated by a merger of Protestant culture and modern science, became synonymous with both his anti–Home Rule stance and the survival of the ether concept.

However, by the 1880s there were indications that the ether concept was becoming scientifically untenable. The American physicist Albert Michelson and a colleague had devised a number of experiments to test Maxwell's suggestion that should the ether exist, then it would be expected to exert a drag on the earth as it moved through ether-filled space. There was no evidence of such a drag and the experiments eventually were deemed to have rendered the ether defunct.[107] In 1889 Fitzgerald was, however, still attempting to reconcile his ether theory with Michelson–Morley's null result, as is clear from a paper submitted to the American journal *Science*, in which he states: "I have read with much interest Messrs Michelson and Morley's wonderfully delicate experiment. . . . Their result seems opposed to other experiments showing that the ether in the air can be carried along only to an inappreciable extent. I would suggest that almost the only hypothesis that can reconcile this opposition is that the length of the material body changes, according as they are moving through the ether or across it."[108] This now famous paper was the impetus for the "Fitzgerald–Lorentz" contraction, which foreshadowed Einstein's thoughts in a 1905 paper titled "The Electromagnetic Properties of Moving

Bodies," in which he introduced the concept of special relativity. Fitzgerald's primary motivation in 1889 was the resuscitation of the ether principle, but by this date various errors and oversights had been found in his formulations, and the sponge theory was deemed unstable by his peers. As Lodge described, "At present the vortex theory of the ether labours under a great many difficulties . . . for it threatens to gradually change its properties with time, containing within itself the seeds of its own entanglement."[109] In 1890, Fitzgerald once again attempted to salvage his theory at the British Association Meeting. Here he demonstrated ways in which the tendency to instability could be delayed and that "the break-up of the sponge would in fact tend to become slower and slower." For the rest of the decade until his untimely death in 1901, he continued to advocate the vortex sponge theory refusing to admit defeat, as he confided to Oliver Heaviside, "I have a sort of feeling in my bones it must be so."[110] The theory thus persisted in his mind as the ideal of what a future unified theory of matter and ether should look like.

The vortex sponge model of the ether was a particularly seductive hypothesis as it offered a theory of the universe that was complete, unified, and fundamentally simple, and it thus remained a methodological guiding principle for many fin-de-siècle physicists.[111] In Dublin, Fitzgerald's reluctance to accept the demise of the ether can be seen in the context of an undermining of what he perceived to be the natural links between modern science and Protestant culture. Dublin physicists based at Trinity College seemed particularly eager to keep the notion alive. Fitzgerald's student and later assistant, Frederick Trouton continued to experiment on the ether until 1908. And as late as 1914, Trouton, who was then president of the mathematics and physical sciences section of the British Association, opened his address with a nostalgic reflection on the golden days of Fitzgerald. He reminded his audience that the principle of relativity "would not have been accepted half a century ago, when a purely dynamical basis was expected for a full explanation of all phenomena" and, he continued, "the men of that generation were the sturdy Protestants of science, to use an analogy, while we of the present day are much more Catholic in our scientific beliefs."[112]

To recap, then, Dublin's elite Protestant scientific community was at odds both with the secularizing professional classes modernizing science in London and with Nationalist scientific philosophies closer to home. Returning now to Fitzgerald's Dublin habitus, the remainder of the chapter airs something of the tensions arising from the shared city spaces that resulted in Fitzgerald's disposition toward establishment traditions over radical change in science and politics. Fitzgerald's problems in promoting science and technology in a predominantly Nationalist city were manifold. First, resistance to any changes propounded by a "self-professed new scientific elite" was strongest among the Catholic hierarchy, who traditionally directed the course of education of their flocks. For their part, and spurred on by the "materialistic" pronouncements of Tyndall, the bishops believed that were it not for their own vigilance, an unbelieving tide would have swept through the entire curriculum. This circumstance justified "to the full the determination of Catholic Ireland not to allow her young men to frequent Universities and Colleges where bishops rebuked the indifference of those who may be tempted to grow slack in the struggle for a Catholic system of education."[113]

Second, Fitzgerald's involvement with a number of government commissions on education left him unhappy with the level of commitment to science that he experienced from administrators. Both he and Barrett played a role in the foundation of the city's first technical colleges and were members of the Provisional Committee assembled in 1887 for the Technical School in Kevin Street, Dublin. An Irish Technical Education Association was founded shortly afterward in 1890 to press for the provision of proper facilities throughout the country.[114] The correspondence between Fitzgerald and Barrett relating to the setting up of the Irish Technical Educational Association, reveals something of the tortuous process involved in this small step toward the provision of technological training in Ireland. Barrett, surmising that the views of this charismatic local scientist would carry weight in London, persuaded Fitzgerald to become involved, despite continued resistance to the plans among educational bodies closer to home. Resigned to a slow pace of change, he admitted to Fitzgerald, "If you knew the inner workings of the present system you would be amazed at its irritating ineptitude and that any educational institution over here could survive the disregard of local opinion that has universally characterised it—an installment of reform is better than none at all and one step leads to another."[115]

As well as sitting on the Board of Technical Instruction, Fitzgerald went on to work as a commissioner for education with the Boards of National Education and Intermediate Education to promote scientific education at all levels of instruction. However, by 1893 his frustrations with the system were still very much in evidence, as can be seen in his correspondence with the boards. In one letter, he laid out his views on the subject of primary school readers and registered his dissatisfaction with the textbooks passed by the board for use in schools. He complained that his attempts to get some readers rewritten were thwarted, and sections that he wished to include had been omitted "because a few sentences have been objected to by Trade Unionists and by Archbishop Walsh."[116]

Finally, his place of employment, Trinity College, found it hard to generate funds for science, a situation that was bound to get worse with the breakup of the Union. Also, any technological aspirations on the part of the science departments were constrained by the perseverance of established academic traditions at the college. Despite being the first of the older universities to attempt to combine engineering with the more traditional liberal arts education, opening its engineering school and courses in experimental science in 1841, the courses remained largely theoretical and graduates struggled to gain employment as they were deemed too inexperienced in practical work.[117] Fitzgerald, having failed to convince the powers that prevailed in the college to support an adequate teaching laboratory, pondered, "When will poor TCD get credit for wanting to do more, very much more than its very limited and precarious income permits?"[118]

In sum, the Catholic hierarchy was erecting barriers to the new science in the classroom, which they perceived as a threat to Catholic education from scientific naturalists. Senior officials on the Board of National Education were unsympathetic to changes recommended by Fitzgerald and Trinity officials lacked the will and money to exact educational changes. The antitechnological feeling at third level, which so frustrated Fitzgerald in his quest to propel Dublin onto the world scientific stage, was percolating down to primary school education, and reluctance to implement changes seemed rife throughout the entire system. Dublin's connection with London was Fitzgerald's only lifeline for a long-term vision for science. Without it, he believed, the scientific potential of new generations of Irish would remain unrealized. Fitzgerald held such strong beliefs in the vital cultural and intellectual value of science that he lambasted proponents of classical education who excluded experimental science from education and left the population lacking in skills. As he put it, "The niggardly recognition of science by the public is a disgrace to the enlightenment of the nineteenth century . . . the public now are but the children of those who murdered Socrates, tolerated the persecution of Galileo, and deserted Columbus."[119] His chagrin extended also to a scientifically apathetic political system, which showed no signs of harnessing the economic potential of science, as he lamented, "Another matter that I cannot endorse is leaving out of the whole subject of political economy [in schools]. It is most important that those who have to govern the country in the future should have some dim ideas of what the questions they deal with are like, and by giving them a few broad principles should be saved from the more obvious fallacies which are the stock in trade of so many political charlatans."[120] These continuous broadsides at an Irish political and educational system, governed by those with "monastic book-wormy habits of thought,"[121] signaled a more drastic intention as he confided in one of his many letters to his colleague Oliver Lodge, "With this Home Rule looming in the near future, I cannot afford

FIGURE 4. The Applied Mechanics Laboratory at the Royal College of Science, Dublin ca. 1920. Such early twentieth-century classes in practical science at the RCSI were the legacy of late Victorian educational reforms campaigned for by George F. Fitzgerald and William F. Barrett. Reproduced by kind permission of UCD Archives UCDA RCSI/248.

to waste more money than I can help, as I shall almost certainly have to leave Ireland if it comes on."[122]

Fitzgerald's arguments for Home Rule drew on Spencerian notions of progressive evolution, when he forewarned that populations lacking in adequate scientific education and without the reasoning skills engendered by experimental science were unlikely to survive, let alone prosper. Home Rule, he implied, would be tantamount to the extinction of the "Irish race." Sounding these notes of caution to an Irish Industrial League meeting in May 1896, he stated, "The way that science bears on industry is that, for efficient work, and especially for improvement, for keeping up with the times, for withstanding the competition of the world, for raising themselves above the position of hewers of wood and drawers of water, for preventing extermination . . . a people must be provided with accurate information and with habits of accurate work, that is with scientific information and scientific methods. I am afraid the people of Ireland have neither the information nor the habits of science." But, of even graver concern to Fitzgerald than the prospects for

Irish industry, was the future of Ireland itself. "Not only does the survival of industry depend on such habits," he warned, "but the very survival of the Irish people, for preventing extermination like that of the Maori and the Red Indian, a people must be provided with accurate information and scientific methods."[123] As David Attis underlines, it was Fitzgerald's contention that the evolution of industry and civilization threatened extinction for those who failed to adapt.[124] Fitzgerald's Dublin cityscape in the 1890s, then, was interwoven with pressing scientific and societal concerns. On the one hand he was attempting to salvage his vortex sponge theory to secure the idea of a continuous universal ether, and on the other he was resisting the dissolution of local political space.

In Fitzgerald's view, the progress and prosperity of Dublin and the survival of its people was dependent on adaptation to the new science, and his solution to problems of politics and science was the same: the connecting links must be attended to and strengthened.

This chapter has demonstrated some of the ways in which the *lived* experiences of William Fletcher Barrett and George Francis Fitzgerald, two physicists working in late Victorian Dublin, helped to shape the spaces of science around them. It has highlighted the role of Dublin cityscapes in the development of their ethereal preoccupation in the late nineteenth century, and the ways their location in regard to this research, whether inside or outside Dublin's established Protestant parameters, influenced the debate on the constitution of the ether, and consequently on the origin of matter. Both practitioners viewed the properties of ether from different standpoints, uncovered its potential in different ways, and were firmly rooted in their ethereal convictions, but were united in mobilizing the *extraordinary complexity* of the ether to underpin their respective doctrines. But, in both cases, whether the ethereal priorities or preoccupations were with the spiritual or the mechanical, the race for definitive accounts or explanatory frameworks, as we have seen, revolved around local considerations.

The appearance of an article in the *Nonconformist* in 1875, by a scientist based at the Royal College of Science in Dublin, espousing the Christian potential of ether can be seen as the beginning of an attempt to shore up public morality against the encroaching tide of unbelief, while another scientist, a short walk down Dawson Street to Trinity College, was mobilizing the same illusive substance to shore up Anglican identity against the rising tide of Home Rule agitation. One was appealing explicitly to the dangers of "materialism," the other found "materialism" idealistically construed, an ally in a battle for unity and political cohesion.

Barrett was driven to protect society from Tyndall's "materialism" and the excesses of commercial life that he saw as providing a united front against the fundamental Christian values of brotherhood and selflessness. The poverty and disharmony he encountered while working in Dublin provided him with the space to promulgate his spiritualistic theories of the ether and at the same time attend to the physical and moral education of the underprivileged. Barrett's "life in science" serves to demonstrate how the modus operandi of this particular habitus helped to shape the institutional and pedagogical spaces around it and to challenge the prevailing climate of science.

Elsewhere in the city, the search for a canonical explanation of the ether for Fitzgerald was an attempt to underpin Protestant spaces in a time of precarious political geography—the very fabric of his life in Dublin was at stake and his dream of advancing the profile of Irish science and technology was threatened by a Nationalist post–Home Rule society. His location and *lived* experiences served to elucidate Edward Soja's conception of "the spatiality of human life" and demonstrate how this "new man of science" was a distinctly spatial entity in a complex relationship with his Dublin cityscape.

I end with a mention of Steven Connor's recent observation that nineteenth-century writings about the ether depended on a kind of dynamic imagination "focussed not so much on how things appear as forms, as on how they worked and felt, as actions and stresses."[125] This resounds well with J. H. Poynting's declaration nearly one hundred years ago that "physicists looked to mechanical models to explain the universe because of the nature of minds, rather than the nature of the universe. . . . [We are] able to think of ourselves as part of the connecting machinery, feeling the stresses and helping to make the strains."[126] These comments, nearly a century apart serve to underscore the contention of this chapter; that "the all-pervading ether" in late Victorian Dublin, was more than a disembodied scientific conundrum, it was a flexible resource being continually molded and reshaped by the contours of scientific lives in the city. The ether forged an important nineteenth-century space for science in which, as Connor configures it, a discourse on the "mentality of matter" and the "materiality of mind" could be played out.

Chapter 2

Form

> Is function the mechanical result of form, or is form merely the manifestation of function or activity? What is the essence of life—organisation or activity?
>
> –**E. S. Russell**

The antiquity of such morphological musings is brought to our attention by numerous studies. First, Theodor Gomperz has pinned nascent stirrings of debate in the field on Empedocles's fourth-century BC observation that "hair and foliage and the thick plumage of birds are one." Johann W. von Goethe, however, earned the epithet "prophet of formal morphology" following his 1807 paper "Metamorphosis of Plants" in which he stated: "This science which should discover the inner meaning of organic *Bildung* is called morphology,"[1] and so the term was coined to denote a search for underlying similarities between different organic forms. D'Arcy Thompson's later translation of Aristotle's treatise *Historia Animalium* presents us with the earliest endeavor to classify and describe the causes of the form and structure of animals.[2] Lynn Nyhart, in an introductory overview of morphology's place in the history of biology, has more recently explained how it ultimately became the science of form, that is, the term now encapsulates the underlying principles governing the organization of natural bodies from animals and plants to rocks, landmasses, and molecules.[3]

The quest to understand form and function was integral to Victorian debates on the origins of life. This chapter focuses specifically on the laws of *organic* form, the establishment and negotiation of which predominated nineteenth-century thought in the life sciences both before and after Darwin's intervention, and tracks the ease and difficulty with which notions of form found refuge in Dublin's spaces of biological investigation at the end of the nineteenth century. The central argument pivots on the diverse attitudes to the problem of form that were expressed in the 1870s by two city-based Presbyterian scientists, David Moore (1808–79) and Alexander Macalister (1844–1919). Moore, a prominent Dublin botanist, was not persuaded by the theory of evolution by natural selection as the mechanism of species transformation, and "his botanical garden in Glasnevin was an imprint of his commitment to the role of God in the design of nature."[4] The coreligionist and Trinity anatomist Macalister, on the other hand, provided Darwinian accounts of anatomical variations, such as vestigial organs, and pursued evolutionary links between human and primate physiology.[5]

For David Moore, Darwin's theory challenged the certitudes on which his life revolved. Moore's defense of natural theology as the dominant motif for explaining the workings of the natural world can be seen, I argue, as an attempt to reassert a worldview in which the "living of his life" made sense. The idea of a naturalistic cosmos of the kind Darwin was suggesting unsettled Moore's convictions considerably. Like the eighteenth-century naturalist John Ray, he believed in a Divine designer who intervened in the creation and adaptation of every organism. Shortly after Moore lost his first wife Sarah, the "only source of my earthly happiness," to typhus, he revealed in a letter to a friend Isabella Morgan that Sarah "gave herself up" while not appearing to "betray any uneasiness about her approaching end."[6] The purposefulness with which Moore believed every act of the Creator was imbued—however incomprehensible to the mortal mind—was reflected in the words of Ray, "Howbeit, I will not deny, but that God doth sometimes Himself immediately inflict diseases, even upon his own children for many good considerations."[7] This theodicy grounded in utilitarian natural theology was the grid from which David Moore read meaning into life's events, from the very personal—the tragedy of his wife's early death—to the acutely public, his duties as director of the Dublin Botanic Gardens. Moore therefore, aided by the workings of his garden and the many facets of a Dublin "botanical life," attempted to publicly reinstate the authority of the utilitarian design argument over the Darwinian model.

On the other hand, for Macalister, whose theological mindset had no qualms about organic transformation per se, the naturalistic implications of Darwin's theory did present a huge challenge to his ideas about the origins of human intellect. The meanings that were attributed to Darwin's theory here were shaped by what Macalister saw as a Darwinian failure to account for the *higher* faculties of the brain, and the consequences thereof for societal development. Late Victorian scientific discourse was, at the time, embracing the views of anthropologists such as Edward Tylor and John Lubbock who were espousing notions of a progressive evolution from primitive to complex societies. In Dublin, however, coteries of Celticity were emerging around the Royal Irish Academy and Trinity College. These Romantic revivalists were calling on historical concepts around successive ages to work out ideas of ancient society rather than on evolutionary dynamics.[8] But more important, these researchers had helped to cultivate the notion that early societies were culturally advanced. The idealist precepts of Macalister's transcendental anatomy fitted better with such ideas than the Darwinian framework of progress that Royal College of Science scholars and Dublin medics at the time were adopting. I suggest that Macalister however, was enabled by his "scientific habitus" to continue his personal research and structuralist agenda while publicly showing allegiance to Darwinian biology. This he did by careful negotiation of morphological principals and rhetorical spaces.

A vast amount of historiographical research in the past forty years or so, has dwelled on the development of both Darwinian and non-Darwinian biological thought throughout the nineteenth century.[9] The discussions have traversed a wide territory, from debates on the mechanisms of evolution and how to interpret them, to questions around form and function, to the social context of the debates themselves. Here, I want to turn to geography, and continue the approach that has already revealed something of the variation in the local reception and mobilization of Darwin's theory around the world.[10] In Dublin, analysis in this area has thus far centered on politically motivated Catholic and Anglican institutional responses to Darwin.[11] This scrutiny of Dublin, however, will delve more conspicuously into the spaces of the city, and ask, among other things, what did engaging with "Darwinism" mean in the context of a metropolitan "scientific life?"

As Dov Ospovat points out, Huxley's reduction of biological explanation to a Creation versus evolution dichotomy in the late nineteenth century has had a disproportionate influence on the historiography of science.[12] In addition to this, running parallel to issues of teleology and biology ran social currents that also have a bearing on discussions. Research in this area has been asking whether Darwinism was fundamentally an idea in science or a broader social philosophy. As David Livingstone observes, "The period was one in which the rationale of social legitimation was transposed from one theodicy, based on natural theology, to another founded on the new laws of nature."[13] Pursued from this particular angle, the interpretations of theories of form by individual scientists can be best apprehended by assessing the social, religious, and cultural mechanisms governing the landscape where the debates were played out.

Moore and Macalister certainly had very different teleological understandings, but to present them solely as examples of creationist or evolutionist Dublin Presbyterians is to miss the wider geographical picture. In recognition of the intrinsic link that existed between nineteenth-century ideas of morphology—"the very soul" of natural history[14]—and contemporary evolutionary debates, the chapter stands back from affirming the rival naturalist stances propounded by Huxley and builds instead on Ospovat's recommendation that we replace Huxley's categories with others that reflect more closely the divisions among naturalists at the time.[15]

So, by focusing on a habitus molded by the mores of Trinity College for Macalister, and one honed by the Dublin Botanic Gardens for Moore, the geographical approach adopted here moves the analysis away from a *codification* of stances, and works instead to reveal the subtle ways in which morphology and "habitus" interacted in late nineteenth-century Dublin.

Kohler reminds us that fieldwork practices depend on the conditions of specific places, requiring considerable improvisation to cope with local exigencies. Thus, by stressing that the field is qualitatively different from the closed and controlled workplace of the laboratory and that scientific work in the field is unmistakably informed by the cultural conventions of ordinary conduct, it becomes clear that the different fieldwork requirements and practices of Moore and Macalister initiated a polarization of their intellectual spaces, a theme that was to continue into the laboratory and beyond. Another spatial consideration to take account of is the fact that Moore and Macalister inhabited "border zones" or "hybrid spaces" between field and laboratory. Both Dublin practitioners embraced the spaces of procurement and acquisition for the purposes of teaching, display, and research, and both experimented, dissected, and sampled in their institutional setting, whether it was a glasshouse or an anatomy room.[16]

The following section examines two themes in its spatial exposition of the lives of David Moore and Alexander Macalister. First, it tends to the biological spaces of structure and function, and here, explores the ways in which Moore's curatorial aspiration to display the beauty of his botanical specimens emphasized adaptation to the "conditions of existence" as the force responsible for determining the anatomical structure of individual species, and how Macalister's emphasis on "structuralism" as the force that constrains the anatomical diversity of life and roots it in a common archetype demonstrated his goal to understand organic form. Second, it draws attention to the ways these morphological differences worked with and against distinct research environments and city spaces, to negotiate Darwinian naturalism.

Maintaining Function at the Dublin Botanic Gardens

David Moore's conception of the relation of function to form ran counter to that of Alexander Macalister. Although both men were philosophically rooted

FIGURE 5. David Moore (1808–1879), director of the Dublin Botanic Gardens. Courtesy of the National Botanic Gardens, Glasnevin.

in Scottish Presbyterianism, Macalister found evolutionary accommodation in Richard Owen's argument that "structuralist biology, properly understood, could contribute towards an understanding of the natural causes of species origins." Moore's utilitarian natural theological explanations of life however, were paradigmatically more resonant of Darwin's thinking, but differed in terms of the causes invoked to explain that form was merely the manifestation of function. In this respect, in answering the question posed by Russell—"What is the essence life—organisation or activity?"—Moore, ironically, as we shall see, had more in common with Charles Darwin and Joseph Hooker than Macalister had.

That David Moore's scientific expertise was lauded in Dublin's academic circles was unquestioned, and none heaped more praise on his talents than the eminent surgeon Dr. William Wilde, who announced to a meeting of the Royal Irish Academy in the early summer of 1860 that "the Royal Dublin Society ought to be proud of having so scientific a gentleman over their beautiful gardens at Glasnevin which afforded pleasure to the public and instruction to all who took an interest in botanical studies."[17] David Moore sought to reclaim authority for a traditional reading of the adaptationist argument when scientists around him were entertaining the idea of natural selection. His functionalist leanings were espoused primarily in a Belfast lecture in 1874—his most vocal promotion of the efficacy of natural theological science—and in the handbook compiled for his botanic gardens, an illustrated sermon on design and adaptation perused by city dwellers for over two decades. Following a brief biographical sketch, the first part of the chapter demonstrates how Moore promoted this functionalist and utilitarian view of biology by a careful negotiation of the public, domestic, religious, and political spaces of his Dublin garden.

David Moore held the post of curator of the botanical gardens at Glasnevin from 1838 to 1870. He was born in Dundee on April 23, 1808, to Charles Muir, a gardener, and Helen Rattray, and received his first botanical tuition from Douglas Gardiner at the local Dundee Rational Institution Museum. Much of his early horticultural training was then accumulated in the gardens of the Earl of Camperdown in Perthshire, where he was apprenticed to the head gardener Mr. Howe. From there he moved to the nursery of James Cunningham at Comely Bank, Edinburgh. In November 1828 he was appointed foreman at Trinity College Botanic Gardens, a six-acre site that abutted on Shelbourne, Landsdowne, and Pembroke Roads in Ballsbridge, Dublin.[18] Here he acted as assistant to James Townsend Mackay (1775–1862), a fellow Scot, who had honed his skills in Edinburgh's Botanic Gardens. In 1832, after the death of Moore's mother, his younger brother Charles, then aged twelve, joined him at the Trinity Gardens as an apprentice.[19] Following an unsuccessful application in 1834 for the curatorship of the Dublin Gardens in Glasnevin, Moore opted for another botanical venture in his adopted country. Under Mackay, he had developed a keen interest in Irish flora, and this in turn had initiated an interest in the biogeography of plants. By good fortune, in June of the same year, he was appointed botanist to the Ordnance Survey of Ireland. With a base now in Belfast, he explored the flora of northeastern Ireland and collected and surveyed indigenous plants from Antrim and Down, where he made and published several discoveries. His excursions in the northwest culminated in the botanical entries to *The Ordnance Survey of the County of Londonderry* (1837).

In October 1838, following Ninian Nevin's resignation, Moore reapplied for the curatorship of the Botanic Gardens in Dublin, and this time was elected. His younger brother Charles replaced him at the Ordnance Survey on his departure. Moore was to have a long and fruitful career at the gardens in Glasnevin, despite occasional conflicts of interest with the Royal Dublin Society.[20] His position was upgraded to director in May 1865, but the Department of Science and Art, which then financed the gardens, withheld sanction until November 1869. Moore married three times and had nine children while in Dublin.[21]

The Dublin Gardens performed both an educational and a scientific function, and with Moore at the helm, improvements were made "of a nature calculated to accommodate the establishment to the present advanced state of botanical science and increase its popularity and usefulness to the public in scientific, educational as well as practical points of view."[22] In keeping with this mission, Moore carried out botanical experiments in the gardens and published widely in scientific journals.

Two significant horticultural advances were made during his time at Glasnevin. First, an avid interest in orchids led to experimental propagation and the germination of species from seed to flowering plant between 1845 and 1849. A submission by him to the *Gardeners' Chronicle* in 1849 recounts "within the last five years, seedlings of the following species have been raised in the Orchid house in Glasnevin, namely *Epidendrum elongatum* and *crassifolium*, *Cattleya forbesii* and *Phaius albus*, the seeds of which all vegetate freely."[23] He subsequently oversaw a program of renewal of the glasshouses for his expanding orchid collection. Second, in the early 1870s, *Sarracenia x moorei*, the first artificial hybrid pitcher plant, was raised and named after him. The hybrids and another achievement, *Aponogeton madagascariensis*, were exhibited at the 1874 International Botanical and Horticultural Congress in Italy, where Moore was awarded two gold medals from the Royal Horticultural Society of Tuscany.

For Moore the visible presence of a design in nature logically required the presence of a designer. He endorsed Paley's argument that every part of every organism had been meticulously designed for its function, and reflected a way of looking at the world that regarded science as the study of Divine laws governing nature. Importantly, Moore believed that the beauty of plants, both their visual appearance and their morphology, expressed the wishes of a designer who sought to provide humans with a pleasing world to inhabit. He interpreted the patterning of flora created in his garden "as testimony of God's intervention rather than as proof of nature's own selection mechanism." For him, like Paley, it was "a happy world after all."[24] On the other hand, "the struggle for existence" promoted a pessimistic outlook, as Darwin himself scribbled: "What a book a Devil's Chaplain might write on the clumsy wasteful blundering low and horridly cruel works of nature!"[25]

This is not to say that Victorian garden curators were innately predisposed toward an optimistic creationist view of the world. As a scientist, Moore's professional spaces would have traversed the same territories as Joseph Hooker at Kew, and yet Hooker embraced Darwin's theory enthusiastically. Instead, local religious, political, and class issues wove their way into Moore's garden, which enabled a teleological understanding of biology to flourish. With his Scottish Presbyterian heritage, he stepped easily into a guiding role in which his theological beliefs were relatively unchallenged by the local constraints of his post.

One of Moore's most emphatic commitments to a utilitarian natural theology was articulated in a Belfast lecture, organized as part of a series titled *Science and Revelation: A Series of Lectures in Reply to the Theories of Tyndall, Huxley, Darwin and Spencer Etc.*[26] Following John Tyndall's provocative opening address to the British Association meeting in Belfast in August 1874, Calvinist attitudes to evolution had hardened.[27] Darwin's theory began to be challenged in Belfast's theological circles, and the pulpits of Belfast were used to refute what were taken to be irreligious claims. The Reverend William Johnson went a step further and set up a series of lectures on the relationship between science and religion to be held at the Presbyterian church on Rosemary Street over the winter of 1874–75. The series included addresses by eight Presbyterian clergymen and one scientist, David Moore. The clergymen looked into the moral, theological, and doctrinal implications of evolution but Moore provided the view of a prominent scientist for whom the claims of natural theology remained important right up to the 1870s.

Although Moore's contribution to the series has already been discussed in terms of its antievolution sentiment in the light of Tyndall's pronouncements in August 1874,[28] something of Moore's attempts to regain scientific authority for his reading of biology, in response to Hooker's address at the British Association Meeting in Belfast, has yet to be assessed. For Moore, evolution by natural selection was simply the wrong way of interpreting adaptation in the bigger utilitarian picture that he shared with Darwin and Hooker, and the lecture provided him with the opportunity to illustrate this. Moore therefore used the address to underpin his "scientific" allegiances with Darwin and Hooker, but at the same time to advance his belief in the superiority of the design version of the utilitarian argument over the neoteric naturalistic argument.

Botanical discourse was abuzz in the 1870s with news of new and exotic insectivorous plants being uncovered by collectors around the globe.[29] Authors as diverse as the art critic John Ruskin and the popular science writer Grant Allen were disseminating views on botany and aesthetics that both incorporated and eschewed Darwinian theories.[30] It was Darwin's report in the summer of 1874 to the Scientific Committee of the Royal Horticultural Society on the insectivorous habit of *Drosera* and *Pinguicula* that prompted speculation on these plants. Much discussion in the botanical and horticultural journals followed. That same summer, Dr. Burton Sanderson had made the claim that in the case of *Dionaea*, "this plant digests exactly the same substances, and in exactly the same way, that the human stomach does." The popular press soon developed an interest in the carnivorous plants and they were said to "have caused so much amazement among readers of the *Graphic* arising from "curious representations of them in that popular periodical."[31]

Moore tells us that for Hooker's address to the British Association in 1874, "he had kept one of the most crowded audiences I ever was present in, almost spellbound during the whole course of his address." Moore's talk too, was delivered to a packed audience: "even the galleries were filled."[32] This gives us something of a sense of the atmosphere surrounding pronouncements, debates, and lectures on insectivorous plants at the time. The race to provide definite answers about these botanical oddities and their internal processes, involved shifting authority between the interpretations of design and unguided nature.

Moore's lecture contained few original observations and it showed that he was keenly aware of the importance of the recently reported findings of Darwin and Hooker. His own correspondence with Darwin had begun the same year.[33] Moore's talk drew heavily on Hooker's address the previous August and, like Hooker, he championed Darwin's botanical research. In regard to the leaf of *Dionaea*, he tells us "most of the foregoing observations have been confirmed by that excellent observer, Mr Darwin, who has for some years had the subject under his investigation."[34]

To begin with, Part 1 drew on the wisdom of John Ray to introduce his functionalist agenda. Moore quoted from Ray's treatise *The Wisdom of God Manifested in the Works of Creation* (1714): "O lord how manifold are Thy works! In wisdom hast Thou made them all." The audience was made aware of the power and wisdom of the Almighty in all spheres—"those not only small, but immensely great—the sun moon and all the heavenly hosts, are effects and proofs of his almighty power"—before Moore retreated to his particular area of choice, and "endeavour[ed] to show pre-concerted design and infinite wisdom in the vegetable kingdom." Parts 2 to 5 of the lecture were all arranged around the advocacy of design and adaptation. His talk, "enlivened and enhanced by beautifully coloured diagrams,"[35] covered design and adaptation in seeds, roots, stems, and leaves of plants, in the morphology of leaves, in the morphology of flowers, and in the fertilization of plants.

Part 2 launched resolutely into the natural theological argument: "We have here, as Dr. Roget observes, a striking instance of that beautiful correspondence which has been established between processes belonging to different departments of nature, and which are made to concur in the production of remote effects, that could never have been accomplished without these pre-concerted and harmonious adjustments." Moore then called on his scientific expertise to underpin special adaptationism: "It is however when we examine their wonderful mechanism that we can form a just appreciation of their adaptations to the functions they are to perform. To the naked eye, a leaf of a tree appears a solid body, contained within a skin or cuticle; but to the vegetable anatomist it is known to consist of two distinct layers, soldered firmly together in life, but after death separable into two halves. Were this not the case, the functions of the elaboration of the sap which circulates through the stems would not be carried on."[36]

Moore devoted a substantial part of the next section to Hooker's talk on the insectivorous habits of *Pinguicula*. He dealt in his response exclusively with the insectivorous genera *Sarracenia*, *Dionaea*, *Cephalotus*, and *Drosera*. Hooker had mentioned Darwin's and Burton Sanderson's results on *Dionaea*. He deferred to Darwin's authority on the matter and spoke of aiding him as best he could in this research with all the facilities he had at Kew. Hooker summed up their findings with the note that "these remarks will I hope lead you to see that though the processes of plant nutrition are in general extremely different from those of animal nutrition . . . yet that the protoplasm of plants is not absolutely prohibited from availing itself of foods such as that by which the protoplasm of animals is nourished; under which point of view these phenomena of carnivorous plants will find their place as one more link in the continuity of nature."[37]

Moore in his riposte challenged Hooker's evolutionary leanings and Darwin's theory, while also acknowledging the valuable empirical work of his evolutionist colleagues:

> Mr Darwin . . . who has found that the leaves of *Dioncea* absorbed particles of muscle and other animal matter, but were insensible to particles of inorganic matter. In this we have indeed a subject for contemplation, and we may well be reverently

> led to ask ourselves, who gave this plant these remarkable propensities? It will be found rather difficult to explain them on the evolution principle. It is however, to be clearly understood that this digestive matter in the plant, dissolving the insects which it captures, is in no way identical with the action of the stomachs of animals in digesting their food; it is only in some way analogous to that operation.[38]

So, although rejecting the naturalistic implications of Hooker's universal protoplasm idea, that all life is materially connected, he acknowledged here the existence of "suffering" in nature. But in such cases, he explained, the beneficent Designer still had humankind's best interests to the fore. The destruction of insects by insectivorous plants could best be understood in theological terms as the result of "the great law of sacrifice which Christianity reveals as the very life of God, and the realizing of which is the highest life of man." Back in Dublin, Moore had noticed in his Orchidaceous and Fern House that hundreds of ants were emptied from the leaf of the Pitcher Plant, but the "honey making bee" seemed exempt from its clutches.[39] This observation had led him to comment that "all the beautiful arrangements and adaptations may be necessary and wisely designed for keeping up certain links in the chain of insect life." In his view, the ants were wisely designed. They nourished the plant that sustained the activities of the "honey making bee," and so were instrumental in the creation of life's "elixir."[40]

In the final part of the lecture Moore spoke of flowers and fertilization. He began with the observation that "the flower, in all its beauty and captivating loveliness, consists only of leaves, changed or morphologised so as to effect most important purposes in the economy of the plant—namely, the production of seed and the continuation of its kind on the face of the earth." He was drawn to both the function and the beauty of the specimens. The stamens and the pistils, while colorful and designed for attracting insects for pollination, were "as well no doubt, as to please and gratify mankind, and to adorn and beautify the external world."[41] He went on to claim that the manner in which the stamens and pistils operate was "surely beyond the realm of chance." Orchids concluded the presentation. These flowers, like the insectivorous plants, had huge popular appeal as well as being of major interest to botanists.[42] Darwin found them useful for demonstrating adaptation for pollination and illuminating the role of sexual reproduction in evolution. Moore concurred with Darwin on most of his ideas, but came to some different conclusions: "The arguments and reasons he [Darwin] brings forward to support his views are certainly ingenious and original. No doubt his theory is correct in the main, but we have seen the flowers of orchids produce seed in our conservatories during the winter months, when flying insects were not visible. We thus see in the important phenomena of reproduction, not only unmistakable instances of design manifested, but indications of vitality, especially when the reproduction of some of the lower tribes of cellular plants is taken into consideration."[43]

In Moore's view, then, where plant adaptation to the environment was uncovered, it was God who provided the means for adaptation rather than any natural developmental processes; thus rather than evolution providing the mechanism for understanding how plants adapt to their environment, it was the Creator who guided the process, "In all nature," he declared, "the simplicity of her works is not less remarkable than their perfection, and their acting harmoniously to accomplish whatever end they are destined for in the boundless conceptions of the Creator." Moore ended his lecture in support of utilitarian natural theology with the benediction: "We shall therefore conclude by again quoting from the sweet psalmist of Israel "oh lord how manifold are thy works! In wisdom hast Thou made them all: the earth is full of Thy riches."[44]

David Moore's spaces in Dublin actively facilitated this "traditional" version of a functionalist reading of botanical science. Paley's notion of Creative beneficence in the natural world was manifested in Moore's garden, which "buzzed with contented life."[45] And he worked steadily—in the scientific arenas of both the field and the study—to maintain this state of affairs by engaging with the "stately," the "exotic," the "beautiful," and "the transcendental" in his day-to-day running of the Glasnevin site. Just as "beauty, ornament and the intricacy of plants and flowers" were the main hinges upon which, as Johnson says, his antievolution argument turned, so too were they the hinges on which his fundamental curatorial concerns of acquisition and display turned.

From the outset, he made regular visits to stately homes and private gardens such as Chatsworth House, and he received many plants from estates around the United Kingdom. But he also developed important links with Botanic Gardens in Europe, India, Ceylon, South Africa, the Caribbean, and Australia. His brother Charles, the director of the Sydney Botanic gardens, who had traveled widely in the South Sea Islands of the New Hebrides, the Queen Charlotte group, the Solomon Islands, and Caledonia "with good results," had furnished the Glasnevin collection with some "beautiful," "rare," and "exotic" specimens.[46] Moore also used the great interest in plants that existed among the estate owners and owners of large gardens in Ireland to expand trial grounds for rare plants not expected to thrive at Glasnevin. The collections at Headford and Fota, for example, attest to this, as does his friendship with Thomas and Janet Acton who provided land at Kilmacurragh to grow *Califuges* plants not suited to the shallow ground and highly alkaline soil at Glasnevin. Here Rhododendrons from Sikkim in the Himalayas flourished from seeds sent by Joseph Hooker. The Actons' garden became an unofficial annex of Glasnevin, growing plants that could not cope with the conditions on the city site. In 1867 Moore could write to Hooker that he saw eleven kinds of his (Hooker's) Rhodedendrons "growing happily at Kilmacurragh."[47]

These professional and personal ties continued throughout Moore's curatorship at Glasnevin and played a large part in augmenting the species range there. Over four decades, he redeveloped the garden, greatly extended the plant collections, established an impressive range of glasshouses, and introduced and cultivated hothouse plants such as tropical orchids. He was also instrumental in the acquisition of the Aquarium house where the prize giant water lily was cultivated.[48] Although imperial collecting and networking activities were mirrored in botanic gardens elsewhere, Moore's acquisitory motivations differed from those of his Darwinian colleagues. It was, for instance, an avid interest in botanical arithmetic that drove Joseph Hooker's collecting policy. His hope to develop plant distribution into a rigorous and profitable study primarily reflected a desire to raise the prestige and status of botany to the ranks of the physical sciences. For Moore, on the other hand, all these networking ventures to diversify and augment the collections were integral to an Edenic concept of the gardens, which—in Paleyite terms—would have it teem with "delighted existence."[49]

Viewing from a broader angle, we see that as the nineteenth century progressed, the desire for botanic gardens gathered steam across Europe and in Europe's colonies and they became key places for representing biogeographical diversity and encouraging scientific debate. These sites flourished as worldwide networks of plant acquisition and exchange, and formed what Bruno Latour has categorized as centers or nodal points in imperial cycles of accumulation and calculation.[50] As Latour describes it, such centers operated around intricate networks of scientific, commercial, and imperial interests, all of which significantly influenced the production and sanctioning of botanical knowledge that emanated from them. But more than this, as Livingstone tells us, botanic gardens were "multifarious spaces; they hankered after The Garden of Eden; they sought to reproduce global biogeography; they exhibited social standing; they wielded biomedical power." These sites have therefore received fresh attention from geographers in recent years.[51] With this in mind, explanations for the persistence of Moore's teleological "special adaptationism" in Dublin's Botanic Gardens is sought here in the interplay of his life-spaces and the "multifarious spaces" of the garden.[52] Drawing broadly on the theme of "communities of function"—the focus of functionalist thought in morphology—Moore is shown to have been driven by a pious *purpose* in which his dual personas as "public servant" and "botanist to the nation" allowed the perpetuation of a creationist ideology in late nineteenth-century Dublin.

Throughout his forty-year tenure Moore's maneuvering of the urban space at Glasnevin crossed a multiplicity of boundaries and incorporated the many relationships afforded by the dissemination of plant knowledge, the exchange of scientific expertise and the creation of garden design, but his final decade saw one spatial consideration that demanded more particular attention. With the gradual loosening of the institutional scientific grip of the Royal Dublin Society, the 1870s saw a greater public role for the garden in Dublin. Now, Moore was entrusted not only with a metropolitan microcosm of nature with which to experiment, observe, conserve, and research but also with the management of a civic amenity. This enhanced public role allowed him to bring his creationist brand of functionalist biology to a wider audience. A concurrent theme of interest to nineteenth-century historians of science, as Bernard Lightman tells us, was the simultaneous transformation of both British society and natural knowledge, which placed scientists in a favored position "to mediate the conflicts generated by changing conceptions of the sources and bases of the social order."[53] So, because botanic gardens were a *means* of displaying and interpreting the natural world, the lives of those charged with their management played a pivotal role in shaping the kind of knowledge that was garnered there. Moore's primary concern, of course, was to communicate evidence for deeper design and the notion of perfect adaptation of an organism to its environment. As the director of the Dublin gardens, he was in a prime position to disseminate this message to those around him.

Historians of science have recently been demonstrating ways in which botany, more than other disciplines, was enjoyed as an edifying pastime as well as a scientific pursuit. This research has concluded that a life in botany enabled greater permeability between the observances of science and a variety of social practices that revolved around the home and community.[54] But this notion of a quasi-feminization or domestication of botany—also termed "sympathetic science"—which softened the distinction between domestic and scientific spaces,[55] can be seen as conducive to Moore's natural theological worldview in a way not shared by Darwinian functionalists. For Moore, there was no limit to the number of ways in which God's handiwork could be appreciated; poetry, watercolor painting, and scientific experimentation all had a place. Unlike Hooker he felt no need to cast botany adrift from its domestic moorings, instead his intention was to maneuver away from Darwinism and advance traditional readings of the functionalist stance.[56]

Moore's explicit celebration of ultimate causes placed him within an outmoded version of the functionalist paradigm until well into the 1870s. Elsewhere, such as at Kew, Darwinian natural selection had taken functionalism on a new route, and visitors to the London garden had already contemplated

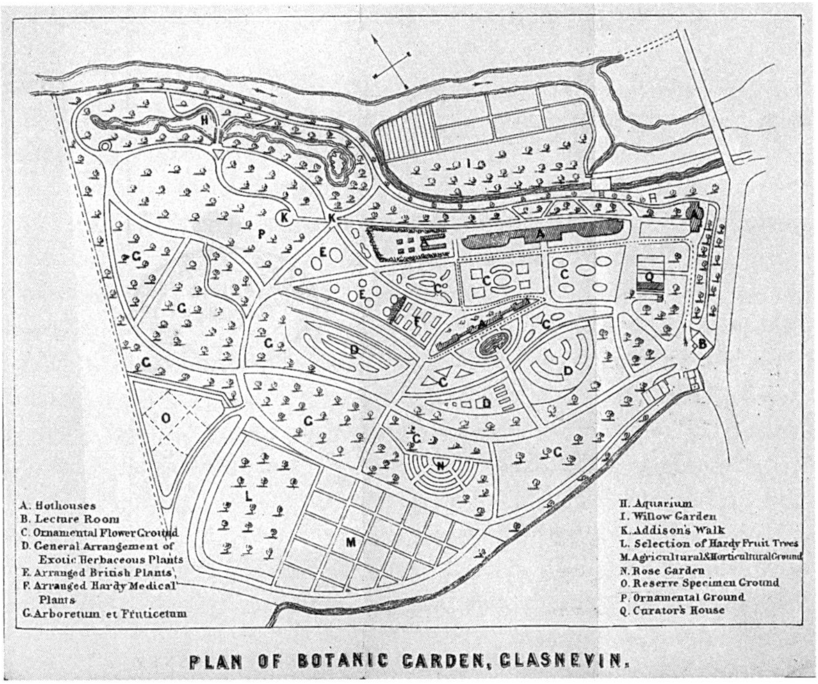

FIGURE 6. Moore's Plan of the Botanic Garden, Glasnevin. This image is reproduced courtesy of the National Library of Ireland. NLI Ir /58072 /m /7.

the notion that species were "derivative and mutable" expounded in publications such as Hooker's "On the Flora of Australia."[57] Not so in Glasnevin. The garden layout in Dublin was a "spatial sermon in Divine design,"[58] and the key to interpreting the lesson lay in Moore's handbook, first published by the Royal Dublin Society in 1865, and reprinted four times throughout the 1870s and 1880s. Here, Moore guided the public through the gardens in a systematic way, which both explicitly and implicitly drew attention to numerous examples of beneficent design and wise adaptation. Intrinsic to the entire publication was a religious subtext that foregrounded natural theology, but also alluded to selective passages of revealed theology in which his botanical specimens featured.

Moore first of all invited visitors to follow his route through the four departments of the garden. The path, he implied, if followed in the manner in which the book instructed, would enable a fuller appreciation of the plants: "If visitors will have the kindness to follow the walks in front of the hot house, indicated by the dotted lines on the annexed plan of the gardens, they will arrive at the respective divisions, according as they are numbered and noticed.

By alluding to this recommendation the different plants will be more easily found."[59]

To begin with, as the handbook takes the visitor on a journey through the flora, it serves not only to provide an education in botanical science and tropical nature for which there was an expanding Victorian market,[60] but Moore, in directing the visitors' gaze initially at the indoor glasshouse collection, was providing a tangible connection with the Divine wisdom of the Creator. Exotic tropical landscapes, particularly palm trees had a major influence on Victorian aesthetic sensibilities, and Moore cleverly sought to enthrall his visitors with the exoticism of his first port of call. Here—where "beautiful species of *Crinums Pancratiums, Hymenocallis and Amaryllis* are frequently in bloom . . . also the beautiful *Bromeliaceous plants*, with Pine-Apple like leaves and rich scarlet bracts enclosing their pretty purple flowers"—senses could be overwhelmed by the "brilliancy of the flowers."[61] And finer sensibilities drawn to the intricate workmanship of plants such as "the lace bark tree of Jamaica, whose inner bark, without any artificial preparation, resembles beautiful lace." Before concluding his indoor tour at the Fern House, the reader is emphatically reminded that "*nothing* can exceed the beauty and variety of the fronds of this type of plant.[62]

Accompanying the doxological tribute of the hothouses were biblical testimonials that alluded to the crucial role of Moore's botanical specimens in illustrating revealed theology. For instance, the Dublin public was guided to the *Papyrus antiquorum* to learn that "the ark in which the infant Moses lay among the sedges of the Nile, which is termed 'the ark of the Bulrushes,' is supposed to have been constructed of this plant." They were encouraged to observe species such as the date palm, or "the true Palm Tree of Scripture," which he explained "affords a good account of the kind of leaves which we are told were strewn before our Saviour when he was entering Jerusalem." The propitiousness of certain trees to the biblical narrative was also noted. The fig tree, *Ficus Sycamorus*, he connected with the biblical accounts of Zaccheus, who had climbed on its branches to view the Saviour, the weeping willow, *Salix babylonica*, which Moore points out "was the tree on which the Children of Israel are supposed to have hung their harps when wandering in captivity," and the Judas Tree, *Circis siliquastrum*, on which Judas was purported to have hanged himself.[63]

More implicitly, Moore drew his reader to particular plants that embodied intimations of Christian iconography in their folklore, for example, *Peristeria elata* the curious "dove" orchid, which had "white marble looking flowers sculptured into the form of a dove in miniature with wings partly expanded as if about to rise in flight." Another was the tree fern, *Cibotium barometz*, also known as the "Vegetable lamb," the rhizomes of which, when unearthed, were covered with long silky hairs. To underpin a more universal appreciation of the associations of nature with the Divine, Moore pauses at the Mexican Hand Tree, *Cheirostemon platanoides*, and tells the reader that according to Humboldt, only one species was known in Mexico, "where it was there held sacred," and he later acknowledges the *Nelumbium speciosium*, sometimes known as the "sacred bean of India," which, he adds, "was still held sacred by the Hindoos, who sculpture the leaves and flowers on the ornamental parts of their temples and other places of worship."[64]

It was of equal concern to Moore on this botanical journey to facilitate an understanding of the beautiful plant forms in terms of the purposes they were intended to fulfill. In the great central house of the range, the Palm Stove, he states, "Palms which are considered the most beautiful family of plants in the whole of the vegetable kingdom. They are also one of the most important to man, yielding him at once as they do, food, drink, material for clothing, domestic utensils and in some instances habitations, their large leaves show how admirably they are adapted to afford grateful shade from the burning rays of a scorching tropical sun; thus further exemplifying 'the wisdom of God in the creation.'"[65]

He drew too on the scientific observations of Alexander von Humboldt penned in the wilderness of South America to cite further examples from the distant native habitats of the Glasnevin plants. In particular, he referred to the flowers of the *Aristolochia cordata*, which the "Indian boys use for caps," and the *Opuntia coccinellifer*, cultivated in Mexico "for feeding the insect which has the power of extracting the juices and converting them to the rich scarlet dye." Regarding the more "curious" and "globular" varieties of the cactus family, he told visitors: "Humboldt informs us that when long droughts occur in the Steppes of South America the numerous herds of horses which roam over these grassy plains frequently save themselves from perishing by scratching away the spines with their feet and sucking the pulp." Moving from the benefits of cochineal, camphor, and cocoa, to the ingenuity of the "vegetable ivory" umbrella handles made from *Phytelephas macrocarpa* seeds, Moore portrays the Palmhouse as the embodiment of the wisdom and beneficence of the Creator.[66]

In its closing statements the handbook invited the public to visit the museum by the garden gate, where they were encouraged to engage with "small articles manufactured from the fibres of curious plants by the natives of foreign countries." Here, Moore had on display "large spathes of the Troolie Palm from Demerara" used by the Huron Indians as Helmets, alongside "Double Cocoa nut from the Seychelles Islands," and samples of Chinese rice paper, all serving as "immutable mobiles,"[67] transposed from the far reaches of the globe to provide a panoptical view of the world for the city dwellers of late nineteenth-century Dublin.

Whereas indoors, Moore's handbook worked to decipher the latent lessons of the hothouses, outside, the scientifically arranged collections were more candidly presented: "The scientific designation of each, with English also as far as can be given, is written on large wooden labels which are placed at the end of each order . . . in such a manner as to make it impossible for the student to get out of one order and into another without knowing," but there remained a sense in which the classification of nature bore the imprint of a higher authority. Taxonomies revealed the regulation of nature, but the visitor—freshly embarked from the exaltations of the Palmhouse, was left in no doubt as to who was ultimately responsible for the orderliness found in the systems of Antoine Laurent de Jussieu, Augustin Pyramus de Candolle, and Carl Linnaeus.[68] For Moore, scientific systems were again evidence of the power and wisdom of the "Great Author of nature," who helpfully bestowed on the student of botany a system of classification with which to read nature's stamp.

As the Moore residence at Glasnevin was embedded in the garden, with family members, students, and employees living on the site, the intermingling of scientific and domestic space was integral to David Moore's botany. For example, the predisposition to systematic detail, as Nuala Johnson notes, extended to "an intense desire to regulate the public's use of the Glasnevin gardens," also permeated the walls of the Keeper's residence, and were reflected in the domestic utterances of Moore to his son, "Even the throwing of your clothes about in your bedroom and on the floor, shewed you had no system in your arrangements." The indistinct boundaries between the spaces of home and science also affected his children's education—insisting that his sons learn Dutch and Latin because "such work requires close attention and above all system in working."[69]

Another example brings to the fore the contested nature of space in botanic gardens, where activities around recreation and investigation had to be carefully negotiated. Moore's eldest son Frederick recalled an incident that incurred the wrath of his father, in which "foes" lurking at the back of one of the glasshouses, happened to be "Edward Madden's giant Himalayan lilies." This "valuable package of seeds," which had journeyed from the mountains of northern India and "had taken many months to germinate,"[70] was ignominiously slashed down with a toy wooden sword fashioned from two pot plant stakes by his children. Moore's concerns in this respect were tested further by the promenading public, leading him to issue instructions to the gate porter to "prevent parties from bringing baskets or other packets with provisions within the gates" and warning that the "police have been directed to prevent persons from leaping over the flowerbeds."[71] Yet, although recreation of an idle or reckless nature was eschewed, leisure in the form of quiet contemplation was actively encouraged.

Because nature study was widely held to encourage sociability and healthy exercise and to promote sensitivity and sympathy, the claim that natural history was morally improving was a common one at the time. As Diarmid Finnegan recounts, the number of natural history and field clubs across Britain in the late nineteenth century reflected the belief that an interest in natural history provided a safe and certain route to moral improvement and physical well-being.[72] But for David Moore, as a practicing natural theologian, this claim had even more relevance. For him, the satisfactory execution of such public duties went hand in hand with a religious worldview in which the Creator had designed his creatures to act in ways beneficial to one another: "Their mutual subserviency to each other, and unanimous conspiring to promote and carry on the public good, are evident demonstrations of His sovereign wisdom."[73] Facilitating the pursuit of local botanical knowledge by providing and maintaining a pedagogic space for the local community, therefore, could be seen in his Paleyite view as accommodating the moral, intellectual, and religious needs of a civil society.

Although a desire for "moral uplift" drove the movement to establish public amenities in European cities, back in Dublin, religious and social considerations meant that this partnership between science and amateur pastime was more nuanced. From its inception the Glasnevin garden had been a contested space where divergent views had made their presence felt. The Royal Dublin Society oversaw the running of the gardens primarily for instruction in scientific botany and "not a promenade for the mere refined amateur." Their critics however argued that "its opening hours were well contrived for the convenience of fashionable and other loungers, and for the exclusion of workers, both the manual and the mental."[74]

The Church of Ireland led the protest against Sunday opening, and proposals opposing their stance had been defeated numerous times in the 1850s.[75] This eventually led to a popular movement on the part of the liberal press, the Catholic clergy, and the "Home Government" to improve access to public spaces for the city's poorest.[76] Surprisingly, David Moore does not seem to have been drawn into the public debate concerning his charge. It would seem, however, from his later correspondence, that any Scottish Presbyterian inclinations he may have had toward opening on the Sabbath were tempered by a number of local factors. First, his Paleyite vision concurred more readily with the Nationalist agitators' claim that "on the continent, by mixing of the classes, good feeling and harmony was promoted, and while the humbler were elevated, the higher did not suffer in position."[77]

Second, he was sensitive to the fact that the plethora of religious arguments being put forward by the Anglican Church and the Royal Dublin Society against "the public profanation of the Lord's Day" were masking a more real concern about the impediment of scientific progress and the perceived "threat to plants" should the gardens be opened to unruly crowds.[78] Moore—who had always foregrounded the "moral benefits" of botanical engagement over any risk of damage and was content that his handbook warnings were sufficient to deter any misdemeanors—had no such worries, claiming that "no injury whatever has been done by the visitors; on the contrary, they appear to take great interest in examining the plants."[79]

He appeared reluctant therefore to align himself with the arguments of the Anglican hierarchy, believing instead that "the excellent effect on the people," which Sunday opening promoted,[80] underpinned the contention that design in nature was most beneficial and that the Deity had added pleasure to the human senses for the purpose of enjoying his manifold Creations. Moore's observations on Sunday visitors were supported by his fellow botanist William Robinson, who described the Sunday crowds attending the gardens: "They come from the odoriferous purlieus of The Combe; from the stews of Church Street; . . . from the many foul alleys on the banks of the Liffey, . . . from all these they come to feast on the beauties of tree, and shrub, and herb, and enjoy the fragrance. . . . Yes, there on Sunday afternoons the 'intelligent workman' . . . may admire [the beauties of the Gardens]."[81]

Following the eventual admission of the public on Sunday, Moore not only was responsible for teaching his audiences how to read and appreciate nature but also was, above all, presiding over a Utopian space. The success of the garden was becoming more dependent on its ability to attract and maintain visitors. The provision of a consistently "inspiring" and "uplifting" venue was tantamount to that success and Moore had risen to this public challenge, noting, importantly, the status of tropical plants in the popular imagination and developing them accordingly. But just as the humid havens of the botanic gardens were propounding notions of an exquisitely designed natural order, the political climate in the city was being reconfigured in the wake of the Home Rule debate. Moore's Paleyite functionalism found space to flourish in this arena also.

As Livingstone tells us, botanical gardens touched on the deepest needs of the state, and our attention is drawn by several authors to the fact that al-

FIGURE 7. The Botanic Gardens Dublin, ca. 1900. This image is reproduced courtesy of the National Library of Ireland. NLI, Lawrence Collection, LROY 8551.

though the idea of a botanic garden as an Eden or earthly Utopia has been long established, in times of political unrest the cultural effects of the idea can become far greater when the site becomes accessible to the general public.[82] Dorinda Outram contends that investment in the "ideal" of the garden as an earthly paradise where the public could experience "moral uplift" by contact with the "ordered display of nature" was partly because this idea of a space "outside space" fitted in well with the new ideas of the "collectivity" or "the nation."

Indigenous botany was being promoted in Ireland as a project of national importance, centrally concerned with questions of national pride and identity, and the way plants and insects had been labeled in scientific surveys of the country was being seen to a great extent as a manifestation of colonial thinking.[83] Natural history's role as a constituent element in this phase of patriotic excitement in the late nineteenth century tied in with Irish linguistic, antiquarian, and folkloric inquiry. Calls from literary revivalists to "arise, botanic Celts and glut your ire!" were being heeded by naturalists such as Robert Lloyd Praeger whose *Irish Topographical Botany* presented plants arranged under localities rather than localities under plants.[84] Others went further and attempted to correlate Gaelic plant names with Linnaean terminology and coin new Irish-language names that followed the Linnaean system to imbue scientific activities with a sense of patriotism.

Because of his posting in Dublin at this time, Moore's botanical spaces were to become inextricably linked to this process of cultural translation. In recognition that collections of a more local kind were beginning to attract public attention in the 1860s and 1870s, his more significant publications of this period were on native Irish flora. His Royal Dublin Society articles, the "Catalogue of the flowering plants and ferns of Dublin and Wicklow" (1878), the "List of the mosses of the counties of Dublin and Wicklow, with their principal localities" (1878), the "List of the Hepaticae which are found in the counties of Dublin and Wicklow, with their principal localities" (1878), and in particular his *Contributions towards a Cybele Hybernica* (1866) compiled with Alexander More, which established the topographical element in Irish botany for the first time, all provided local distributions of plant species and frameworks for subsequent regional floras that dealt with situated botanical knowledge.

Just as Moore in his garden guidebook "encouraged visitors to see beneath the surface arrangements for deeper design,"[85] so too were they encouraged to appreciate the indigenous and the vernacular. Here we see that Moore the botanist was contributing to the spaces of a multifaceted cultural revival in which natural historians played an explicit part in the establishment of Ireland as a biological and geographical territory with an identity of its own. The site at Glasnevin was a space for the investigation of order in the natural world. For Moore "the structuring of the scientific garden and the botanical discourse attending plant life there revealed the workings of a beneficent designer and thus was a material expression of natural theology, the classifying of plants into families, the orderly fashioning of the beds, the display of exotics in the hothouses all facilitated a particular reading of designed nature which confirmed his commitment to a Divine designer." But Moore's system was conducive to numerous social and political interpretations as well as a Divine one. For most of the visiting Dublin public this same functionalist language and utilitarian idealism, which rendered the order of nature visible, accessible, and transparent to the people, also reaffirmed a sense of natural order. The botanical juxtaposition of the exotic with the local, the profusion of familial frameworks, and the celebration of native species, all worked to heighten territorial loyalties and patriotic sensibilities. For Moore, the Dublin gardens, as Johnson tells us, provided a "mediating link between his private convictions as a practising Presbyterian and his public role as a working horticulturalist and botanist."[86] But being based in Dublin allowed Moore space, not to embrace post-*Origin* functionalism in the form of Darwinism, as Joseph Hooker at Kew had done. Instead, his Cuvierian functionalist display of botany in late nineteenth-century Dublin not only promoted civic virtue but simultaneously evoked "the Divine" in a Scottish theological reading and inspired "nationhood" in an Irish political reading, and thus maintained for Moore a currency and relevance for his ideological and scientific spaces in Glasnevin, which elsewhere may not have flourished.

Finding Space for Structuralism in the Trinity Anatomy School

The previous section referred to the notion of "sympathetic science" as an aid to disclosing more about the overlapping domains of the practices of botanical science and the "living" of a botanical life. It did this by probing some of the community networks that emerge around botanical scholarship and by attending to charitable "feelings" promoted by biological investigation operating in the public sphere. Different emotions surface when we examine scientific pronouncements that were forged in darker corners and more peripheral city spaces, rather than those constituted, as in the case of David Moore, at Dublin "beauty spots" and public focal points. I turn now to the spaces congregating around the practices and procedures of anatomy, which inhabited a much less welcome territory for communal engagement.

Here the idea of "location and locution" is mined more specifically. As Livingstone has argued, "The dynamic reciprocity of the spatial and the social mean that there are "spaces of speech" where what is said or heard is mediated by socially and spatially constructed norms.[87] What could and could not be stated *socially* about biological research in the nineteenth century varied enormously depending on the communicative opportunities and constraints of the branch of science in question. The sweep of a broad brushstroke is needed here to introduce key spaces pertinent to the discussion before drawing attention to the ways in which Macalister's rhetoric around morphology, like Moore's, was shaped by these local considerations. I suggest that Macalister, however, was enabled by his "scientific habitus" to continue a personal structuralist agenda while publicly showing allegiance to Darwinian biology. This he did by careful negotiation of morphological principles and rhetorical spaces.

First, botanical discourse was for the most part a suitable activity for mixed social gatherings—prompting, at worst, debate on the merits of materialist aesthetics over the influential views of John Ruskin[88]—but talk of anatomical investigation was taboo, as we shall see. Family members were included in the cutting and pruning activities of gardens and horticultural explorations, but they were shunned from the trials of the dissecting room. This partition of family continued into the professional arena also. As botany acquiesced broadly to its feminizing grip in the nineteenth century, anatomy vigorously demarcated the gender spheres, and in concurrence with the view "that it was an indelicate thing for the young ladies to mix with other students in the dissecting room and lecture-theatre," Trinity College arranged for a separate laboratory to be built for female anatomy students at a cost of £1,500.[89] These detached laboratories and reading rooms also became havens for women to socialize. Tea in the ladies room of Cecilia Street medical school, for instance, was described by one female student as the "most pleasant experience," where they "dissected nothing more gruesome than plum cake." This was in contrast to the plight brought to the attention of the *National Student* of hundreds of male students who had to put up with "the hall, street and neighbouring taverns" to fraternize between lessons.[90]

Just as the "speech spaces" of the anatomical laboratory were unavailable to female students, the siting of these buildings also conditioned conversations and sanctioned the withholding of knowledge to all but a few specialists. Whereas the botanist David Moore was engaged in an open and aspirational city space, and his objects of study were deemed by contemporary accounts to be "beautiful" and "exotic," Macalister's work was hidden from view in a building once condemned as a "public nuisance." The Trinity School of Anatomy, erected in 1825, was strategically located at the very back of the main college "on a marshy piece of land" two sides of which were "lanes inhabited by a very low and disreputable population." A stone wall was built to separate the School of Anatomy from the College Park, and the only access to the main college was via a gate that was kept permanently locked. The professor of anatomy, James Macartney, protested the plans, citing "the demoralising influence of the local environment" where the only entrance permitted was by means of a lane "where there is a gambling pothouse, in which several of the students have been scandalously demoralised even during the day."[91] Forty years later, anatomy was still considered a potentially demoralizing science, according to a General Medical Council Report on education in 1869.[92] The report noted that for many medical students dissection was a "disagreeable duty" that many of them shirked, and only the most diligent persevered with the "appalling work."[93] The experience of dissection, promoted by the profession as a mark of proper medical training, was subsequently perceived by some of its members as having led to "deterioration of character" in their colleagues, by "destroying" a proper regard for the decencies of life.

But if dissecting spaces were deemed to denigrate the character of the anatomist, the fieldwork spaces fared no better. In marked contrast to Moore's "stately" visits, the fieldwork of those in Macalister's profession inhabited much less salubrious spaces. Barely visible from Moore's urban "Elysium" were the watchtowers rising from the adjoining cemetery at Glasnevin, which, until the passing of the Anatomy Act of 1832, bore witness to a very different space of procurement and acquisition. The towers were erected during the nineteenth century to deter body snatchers.[94] These "resurrectionists" or "sack 'em ups" made a living from stealing fresh corpses and selling them to the medical profession for dissection. The death rate among the pauper inhabitants of Dublin city in the mid-nineteenth century was very high, and this illegal and indiscriminate trade was a major source of distress for citizens. Adjoining the grounds of the Royal Hospital at Kilmainham, was a free and unguarded burial ground known as "Bully's acre"—referred to by Macalister as the "Aceldama of Dublin for a thousand years"[95]—which had long kept the resurrectionists in employment.

The Anatomy Act of 1832 was passed, as Ruth Richardson tells us, to protect respectable families from becoming victims of grave robbing for profit; to put a stop to cadaver trafficking and murder; and to increase anatomical cadaver supplies from legally authorized sources. However, the new legislation, in an ironic twist, exacerbated trafficking activities.[96] Pauper cadaver acquisition "now operated through a greater diversity of "official" channels, notably asylums and Poor Law Unions."[97] The specifics of the Anatomy Act now made it easier for medics to "legally" acquire bodies, and an underresourced Anatomy Inspectorate increasingly became embroiled in unlawful and unethical ways of obtaining and distributing corpses for anatomical examination.[98] Unsurprisingly, "the anatomist" became a figure of fear in the community. Dissection practices contravened social mores about how corpses should be treated after death, and it became a commonly held view among the public that the anatomist was promoting the desecration of the body and was therefore engaged in an assault on the soul. Dissection was also associated with crime and punishment, and conjured memories of earlier spectacles of anatomical theaters and crimes that were "punishable by dissection."[99] Unlike Moore, the research material for Macalister's profession was not readily forthcoming from generous public donation.

So the habitus of a city anatomist had to be fully tuned to the spatial sensitivities of place, and rhetorical and textual strategies became part of the armory to deal smoothly with the exigencies of cadaver acquisition and dissection. Failure to negotiate these public spaces could lead to public humiliation, as Robert Knox found to his cost in 1820s Edinburgh,[100] or to messy public confrontations, as in the case of Trinity's School of Anatomy and members of Dublin's Humane Society of St. John. The society's wish was "to protect the

FIGURE 8. Alexander Macalister (1844–1919), professor of anatomy and chirgury at Trinity College Dublin. Welcome Collection V0026757 CC BY 4.0, https://wellcomecollection.org/works.

remains of our poor fellow citizens," part of which involved organizing guards for burial grounds around the city. In a letter to its committee, James Macartney, the professor of anatomy at Trinity, claimed that the organization's charitable work was impeding medical science and he pointed to the "twofold evil effect the society was likely to produce." He continued more provocatively in a newspaper letter, "Nor does it seem wise to discountenance the practice by which many of them are supplied with artificial teeth and hair."[101] However, by the 1880s Macalister had fine-tuned the language of the anatomist to such an extent that when the economic interests of asylums and guardians of the poor converged with the research and training agendas of the Cambridge Anatomy School, he wrote persuasively to the Poor Law Unions speaking of their "public duty," and of the "financial sense" of cooperation, before glossing over the topic of body dismemberment, and issuing false assurances about "the fate of corpses."[102] So, in contrast to the uplifting aspirations of Moore, Macalister's laboratory and fieldwork spaces were grounded in the all too mundane, bordering the shadowy spaces of corpse trafficking, and inextricably bound up with another side of late Victorian city life, which revolved around poverty, disease, and death. His public preoccupations were far from a celebration of the harmony found in nature, they centered, instead, on the circumnavigation of discord.

On a more philosophical note, Macalister, although holding personal beliefs about transcendental claims of the kind Moore alluded to, was not so readily accepting in his public pronouncements of a Divine interpretation for the detailed mechanisms of ordered structure, and was convinced that the "shred of old prejudice against researches in the principles of biology" that continued to be entertained by some religious partisans was the preserve of the "narrow minded."[103] His intellectual preoccupations with "variations," "anomalies," and the "vestigial" seemingly directed him toward Darwinian spaces in search of explanatory frameworks for the less beautiful, the apparently purposeless and hidden aspects of biological life. This journey should have been a straightforward one in the evolutionary hub that was the Trinity Anatomy Department at the time.[104] However, unlike Moore, Macalister's working spaces, although not conducive to a celebration of his particular brand of theism, were helpful in the development of a set of complex rhetorical strategies to achieve his goals. Before attending then to his time and influences at Trinity in more detail, a brief biographical sketch is in order.

As David Moore's long career at the helm in Glasnevin was drawing to an end in the late 1870s, Alexander Macalister's was just beginning. Macalister's family had come to Dublin from Paisley, Argyllshire, in the early 1840s. On arrival his father took up the post of secretary to the Sunday School Society of Ireland. Alexander Macalister was born on April 9, 1844, and Robert Macalister "blessed with a large family and slender purse" destined his son to a business calling.[105] However, as a schoolboy Alexander displayed a love for biological study and spent all his available leisure time at the Botanic Gardens in Glasnevin. Here he attracted the interest of David Moore who persuaded Alexander's father to let him pursue a scientific career. Macalister subsequently commenced his medical studies at the Royal College of Surgeons of Ireland at the early age of fourteen. He was appointed as demonstrator of anatomy there a year later while at the same time studying for a medical degree at Trinity College Dublin. In 1869 he was elected professor of zoology in Trinity, and in 1872 took up the newly created post of professor of comparative anatomy at the college. He was finally appointed professor of anatomy and chirurgery at Trinity College and surgeon to Sir Patrick Dun's hospital, Dublin, in 1879, a post he held for six years. During this time, Macalister was also an assistant surgeon at the Adelaide Hospital, an honorary professor of anatomy at the Royal Dublin Society and one of the secretaries of the Royal Geological Society of Ireland, of which he became president in 1873. He married Elizabeth Stewart in 1866, and had two sons, Alexander Stewart Macalister and Hugh Kidd Macalister. The family lived in Rathmines, Dublin.[106] In addition to his professorial duties, Macalister lectured on botany, geology, and astronomy at Alexandra College Dublin. While still in his early twenties he was reported to have taken part in over 690 dissections, and between 1864 and 1883 he published a large quantity of papers, monographs, and books on animal morphology and human anatomy.[107] This output included two large treatises on zoology titled *An Introduction to Animal Morphology and Systematic Zoology* (1876) and *An Introduction to the Systematic Zoology and Morphology of Vertebrate Animals* (1878) and although each was again "packed with a vast accumulation of facts," it was pointed out that the facts were "unconnected together by any theory pervading the whole."[108] This was to become a common theme in critiques of Macalister's work by colleagues.

Macalister was a member of the senate of Trinity College Dublin and received an honorary doctorate there in 1892. He was elected a fellow of the Royal Society and served on its council between 1894 and 1895.[109] In 1887 he became a founding member of the Anatomical Society of Great Britain and Ireland, later a fellow, and ultimately president in 1898, prompting the observation that "the anatomical Society must regard Macalister as one of its parents."[110] A Royal Society obituarist interpreted matters slightly differently: "No one who remembers his presidency of the Anatomical Society is likely to forget the impromptu references to anatomical literature with which he *overwhelmed* youthful investigators,"[111] and an unsigned obituarist in *Nature* claimed that "Professor Macalister never seemed to realise the crushing effects of his vast erudition."[112] Macalister was also a fellow of the Society of Anatomists of Ireland and a member of the Royal Irish Academy serving on its council in 1874–76 and again in 1880–84. He died in Cambridge in 1919 having held the chair of anatomy there for thirty-six years.

Macalister's numerous biographical accounts and obituaries picture an energetic man, "endowed with a marvellous and orderly memory," who elevated anatomy from "a mechanical study into a living science." However, we are continually alerted to his reluctance to comment on "the problems of biology" and "to the endless pains and infinite patience in obtaining and noting new facts," but unusually, "he could rarely be prevailed upon to publish them."[113] In an article on the muscular system of bats in 1872, Macalister himself states that "the author has for purposes of brevity, carefully abstained from adding anything of a theoretical deduction to this paper, which he has endeavoured to confine to a simple statement of facts." It was to avoid the dangers of "hasty theorizing" and to "discourage the use of the scientific imagination" that Macalister took such actions according to one commentator, who went on to point out that "this thirst for facts and lack of interest in their interpretation is nowhere more clearly displayed than in his choice of subjects for investigation, perhaps the most striking incidence of this is his monograph on the lachrymal bone, a tiny fragile facial component of little interest to contemporary biological debates." According to his *Royal Society* obituary, Macalister's attitude toward the problems of biology was one of restraint, as though he were "continually repressing a natural tendency frankly to probe the real meaning of things, but was held back by the formula he had adopted as the guiding principle of his life."[114] Accounts of this sort portray Macalister as a careful scientist whose religious beliefs prevented him from engaging fully with contemporary biological debates, these accounts also point to a compromise between his early religious training and his scientific experience.[115] Instead, this chapter suggests that Macalister's practices and pronouncements on anatomy can be best understood not in terms of a personal conflict between scientific and religious beliefs, but in terms of his swimming against the reemerging tide of functionalist thought in late nineteenth-century biology, which loomed in line with Victorian ideas of progress.

Despite his stated evolutionism, we know from his writings that Macalister was not an outright advocate of natural selection. He was instead searching for an all-encompassing evolutionary scheme that might eventually see a way to incorporating human psychical development. This scheme would also have to take into account the *duality* of man's psychical nature, which, calling on Schelling's definition, Macalister describes as "the distinction of the soul as thinking, cognition, or idealistic momentum, as discriminated from the combination of sensibility and irritability which make up instinct."[116] Support for his scheme lay with Darwin's research. The mechanisms of natural selection, Macalister concluded, failed to account for the "derivability of man's soul from the psyche of the brute."[117] So, as he saw it, Macalister's functionalist options were limited. If Darwin's theory left the origin of life to the chance occurrence of individual variations and environmental conditions, and Jean-Baptiste Lamarck's had a built-in progressive force behind the transmutation of species, then the only possible paradigm within which he could advance his biological ideas was that encompassed by "structuralism," where the derivation of species was understood to "recognize a purpose in the defined and preordained course, due to innate capacity or power of change."[118]

He thus maintained an avid interest in evolutionary schemes that eschewed the utilitarian nature of Darwin's idea, such as the theories of Richard Owen. Owen, although originally opposed to transmutation, eventually did believe in the efficacy of secondary causes. Owen's position on species origins as outlined in *On the Nature of Limbs* states:

> To what natural laws or secondary causes the orderly succession and progression of such organic phenomena may have been committed, we are as yet ignorant. But if, without derogation of the Divine power, we may conceive the existence of such ministers, and personify them by the term "nature" we learn from the past history of our globe that she has advanced with slow and stately steps, guided by the archetypal light, amidst the wreck of the worlds, from the first embodiment of the Vertebrate idea under its old Ichthyic vestment, until it became arrayed in the glorious garb of the human form.[119]

Here we see that Owen's understanding assumes the existence of natural laws in the formation of new organic forms, but as far as Macalister was concerned, Owen had demonstrated that natural causes need not mean Lamarckian transmutation or Darwinian natural selection.[120] In fact Macalister's publications show that his thinking was in full agreement with Owen's conclusion that "being unable to accept the volitional hypothesis, or that of impulse from within, or the selective force exerted by outward circumstances, I deem an innate tendency to deviate from parental type, operating through periods of adequate duration, to be the most probable nature, or way of operation of the secondary law, whereby species have been derived one from the other."[121] Macalister struggled to promote these ideas in his scientific spaces, but did find support in Dublin archaeological, historical, and theological circles, and equally, his research in these areas strengthened his commitment to Owenite biology. Ironically, the survival of Macalister's evolutionary thesis in

Dublin was dependent on research beyond the bounds of science. As one of his obituarists, Grafton Elliot Smith, recounts, Macalister—whose "skill with the knife was equalled by a strange combination of modesty and pride in his amazing feats of memory"—used these talents in the pursuit of knowledge beyond the lab, and cultivated hobbies such as Egyptian archaeology, philology, and paleography.[122] By 1887 he was publishing hieroglyphic translations from the Fitzwilliam Museum, but already while at Trinity in the 1870s, his writings contained quotations from Greek and Latin writers and references to Hebrew and Egyptian literature and archaeology. At this time too, Macalister had developed a friendship with Peter le Page Renouf, Newman's fellow Tractarian at Oxford and a professor of ancient history and Oriental languages at the Catholic University of Ireland with whom he shared an abiding interest in Egyptology. Together, they studied the ancient Egyptian resources of the Trinity College Library.[123] The conversations enabled by these extramural spaces allowed Macalister to engage in wide-ranging debates around the development of human society. The issues here were twofold: the first was concerned with the question of whether or not a fundamental distinction existed between man and animal or they were simply different levels on an evolutionary ladder; and the second issue, building on this, was concerned with the dynamics of cultural progress. Such exchanges would feed into and nourish Macalister's biological ruminations, as we shall see.

In terms of cultural progress, by the late 1860s and early 1870s, anthropologists such as Edward Tylor (1832–1917) and the X club naturalist John Lubbock (1834–1913) had begun circulating notions of a Lamarckian progressive evolution from primitive to complex societies.[124] Lubbock had emphasized what he perceived to be the immorality of "savage modern tribes," and assumed their behavior was characteristic of Stone Age humans. Tylor had described this early form of behavior, which he termed "animism," not as immoral but as a type of primitive rationality that became more sophisticated over time. These evolutionary anthropologists had built on Herbert Spencer's ideas propounded in *Social Statics*, where he had drawn associations between free enterprise capitalism and individual gain, equated goodness with success, and held the external world responsible for human actions over internal moral values, coining the phrase "survival of the fittest."[125]

It was the implication of these theories, encompassed by Spencer's statement that there is no difference between the rationality of man and that of animals, however, that drew Macalister into the debate. He pointed out that in order to prove such theories, "We require to find gradations in the manifestations and qualities possessed by the psyche of man and animals to find traces of a rudimental or lower soul-form in primitive man, and to see traces of its similar gradual development in the lines of animal descent parallel to that of man's." However, the presence of lower soul-forms in primitive man was not something that Macalister considered would be revealed empirically due to his religious conviction that the soul was "a special Creation bestowed upon him directly by his Creator."[126] Fortunately for Macalister's stance, however, ideas conducive to his reasoning were circulating around the Royal Irish Academy and Trinity College at the time. These were expounding notions of ancient society that were being worked out on historical concepts around successive ages rather than on Spencerian evolutionary dynamics. During Macalister's Trinity tenure, he allied himself with emerging notions of Celtic identity advanced by antiquarians such as George Petrie and Samuel Ferguson, and newer research into local ancient history and archaeology began to hold his attention. Here, as the late prehistoric Iron Age was being increasingly represented as a heroic era, the discoveries of exquisite metalwork such as the Ardagh Chalice in 1868 gave added impulse to the depiction of early Christian Ireland as a "Golden Age" of learning, sanctity, and artistic achievement.

Petrie's and Ferguson's studies of Irish prehistory had helped to cultivate a nostalgic romanticism for ancient Ireland, and had built on the growing Celtic revivalist mood among literary figures that early societies were culturally advanced.[127] To these scholars the Irish past may have been simple in a material sense but was certainly not morally or culturally unsophisticated: "Our gold crowns, collars, bracelets, anklets, our brazen swords, spears and domestic vessels—our cinerary urns, our cairns with sepulchral chambers . . . these are the evidences of the early colonization of Ireland by a civilised people."[128] As Greta Jones points out, ideas of Victorian progressive evolution were being contested by claims such as these, in late nineteenth-century Ireland.[129]

Thus, Macalister, sufficiently buoyed by his particular Dublin experiences, argued against the progressive theories of Darwin, Lubbock, and Tylor by stating that man's morality and religious belief "has no correlate in the lower animals" and that any supposed evidence of an absence of these qualities in primitive societies, was not enough proof of the absence of belief because, based on his research into ethnology and philology and the archaeological finds of his colleagues, he could lob the rejoinder that "races, the lineal descendants of the great nations of antiquity, still live, and yet how many of the ancient useful and ornamental arts possessed by them are lost."[130] What the antiquarian authors were advocating in Dublin was not a common evolutionary path, but separate and distinct civilizations based on racial individuality. Societies, they believed, might be drawn into the modern world, but as Greta Jones explains, the convergence between them was not an inevitable expression of evolutionary law but rather the unfortunate consequence of historical conquest. Here we can see how Macalister's interest in the biological possibilities of structuralism may have worked in tandem with cultural ideas being propounded by colleagues around the possibility that "Celtic and Anglo Saxon civilisations could once again drift apart, separated by their different temperamental and spiritual characteristics."[131] This was in complete contrast to the Lamarckian and social Darwinian progressive evolutionary model that placed the Celt and Anglo-Saxon at different levels of a hierarchical ladder.

Back in the Trinity Laboratory for a time, Macalister too had some respite from progressive evolutionism. His scientific position was in part shaped by his Trinity colleagues and forebears, and it embraced evolution in the context of Berkeleyan and Kantian metaphysics as seen with figures such as William Rowan Hamilton, George Fitzgerald, Charles D'Arcy, and, in particular, Samuel Haughton. In these evolutionary spaces, natural selection was not the accepted mechanism of change but Owen's belief in the derivation of species, which recognized "a purpose in the defined and preordained course, due to innate capacity or power of change," held sway.

During the 1870s—the years that represent Macalister's period of greatest activity as an anatomist—his main mentor and influence was Samuel Haughton, a professor of geology at Trinity and an ordained clergyman. Haughton was a known follower of Owen, and interpreted the fossil record as exhibiting stages toward the progressive realization, by successive creation, of the perfect archetype, thus promoting its Platonic credentials. In Trinity, Macalister and Haughton developed a friendship that had a far-reaching influence on Macalister's career. It introduced him to "something more than the portals of the University of Dublin, for it determined the character of his work and played a large part in shaping his outlook."[132] Haughton had become convinced that the very different modes of organization of limbs and muscles in the wide range of animal species that he had examined in detail could all be explained in terms of one common principle. This was the principle of least action, which he had taken from mathematical physics where it plays a fundamental role. Haughton maintained that the least action principle was not compatible with Darwinian evolution and was convinced that he had made a strong case for the rejection of Darwin's theory. While investigating the problems of animal mechanics Haughton had secured control of supplies of vertebrate material from the Dublin Zoological Gardens and provided Macalister not only with advice but also with the raw material needed for his work.

Other Dublin influences on Macalister's biological thought, which continued in this vein of opposition to natural selection, emanated from a little-known Dublin medic named Henry Freke, who produced and published a number of evolutionary works at a house on Holles Street, Dublin. Freke was a professor of medicine at Dr. Steevens' hospital and "claimed to have established by induction the theory of which Mr Darwin had attained by analogical reasoning."[133] Henry Freke, however, had made no reference to natural selection in his endeavors. Instead he stated that "there exists a chain of progressively advancing organising agents or atoms, whose function is to confer organisation . . . we believe in an equal necessity before that function can be fulfilled, for the existence of a corresponding chain of progressively advancing organized structures whose function is to receive organisation, and which were designed and adapted for calling into operation the function of the organising atoms."[134] Macalister commented that Freke deserved credit for many original ideas and his works on organization deserve to be "closely studied" and if "to the important theoretical ideas put forward in it, he had added any practical confirmatory observations we have no doubt but that his book would have long since taken a foremost place in physiological science."[135] So, Macalister found himself in a position whereby the new science of Darwinism had not pursued the structuralist evolutionary program that Owen had hinted at in 1849. Haughton's work, which he advocated, was being superseded by a new generation of biologists in Dublin, and Darwin himself had dismissed Freke's "interesting" ideas."[136]

Unfortunately for Macalister, the interpretation of the vertebrate archetype as a "metaphysical entity" had booted Owen's morphological work to the sidelines in the nineteenth-century march of science toward natural causes.[137] By the time Macalister took up his Trinity appointment, Huxley and Spencer had embarked on a campaign to discredit Owen's scientific preeminence by attacking transcendentalists' theories, and a number of their colleagues had been strategically placed in Dublin scientific establishments to promote the cause of scientific naturalism. However, in the intellectual climate of Trinity College, Macalister found space to continue with Owen's program for a time. He appeared to pursue the comprehensive empirical investigation of anatomical form left undone by Freke, believing that "the question of development must be decided by work in the dissecting room, the field, the zoological garden and the laboratory; and no authors can expect to receive credit for their opinions but such as the eye, the scalpel, the microscope, or the test-tube can demonstrate what they assert, or, at least the existence of a tangible ground for their assertion."[138]

Macalister also negotiated some of the "communicative" spaces that facilitated biological debate in Dublin in other ways. Being mindful here also of the habitus of the anatomist, in which the procedures and practices of the profession, as we have seen, were imbued with speech and text spaces in which "every act of communication exclude[d] as much as it include[d]."[139] I suggest that the rhetorical devices that Macalister used to maintain his scientific credibility in Dublin had been honed in an anatomist's habitus that was disposed toward the circumnavigation of contentious spaces.

Macalister's pronouncements on evolution while at Trinity reflect Owen's thoughts on species origins as outlined in *On the Nature of Limbs*, where he assumed the existence of secondary causes in the formation of new organic forms and believed that the study of homology would give new insights into those laws. Although Macalister was a structuralist like Owen, his language is staged to befit a multitude of evolutionary textual spaces. It must be remembered therefore that when Macalister spoke of and promoted the evolutionary viewpoint, he was referring to a particular kind of non-Darwinian and non-Lamarckian development, but he did this in terms couched to keep such readers on side. In the much-acclaimed student textbook *An Introduction to Animal Morphology and Systematic Zoology* (1876), Macalister laid down his Owenite argument, "While explaining the method, natural selection does not explain the cause. In artificial selection the cause is the presidence and direction of human intellect. But in natural selection there is a necessity for predicating the existence of a presidence similar in kind, but grander in degree as the changes affected by it are greater than those that artificial selection can accomplish." He goes on to say, "This evolution theory explains the teleological correspondences noticed in animals, for by hypothesis the organ progresses according as the necessity for the function increases. Evolution also is morphological, that is, structures progress only along certain lines, and of this we can give no explanation, but can only say that it has been so, that is, evolution has progressed under law."[140] Two years later in his 1871 critical review of Darwin's *The Descent of Man* in the *Dublin Quarterly Journal of Medical Science*, a journal that had on its list of editors prominent Dublin Darwinians such as William Wilde, he chose his topics for discussion carefully as the new generation of materially minded scientific professionals had begun to take over the control of scientific definition. Here with an eye on what passes as acceptable science in a Dublin medical journal, he outlined the areas of anatomy "from which supporters of evolutionary theory draw their strongest arguments i.e. 'Rudimental structures,' 'variations leading from the human to the simious type,' and 'embryology.'" He concluded that "the defenders of evolution have the best of the evidence on their side," and he discounted any teleological argument for these developments. In regard to rudimental organs, he claimed, "Now how to account for these rudiments on any other but an evolution theory it is very hard to see. No teleological reason for their existence can be given, as they are for no end, perform no function; we can otherwise give no intelligible reason for their presence." In regard to embryonic characters, he says, "There is not one which can be accounted for on teleological grounds."[141] Macalister goes as far as to admit that the evolutionists hold a very strong, if not unassailable position in their arguments regarding the human corporeal frame. Here we see that he employs the use of umbrella terms such as "evolutionary theory" and "evolutionists" to keep Darwinists onside while continuing to work within the framework outlined by Owen in *On the Nature of Limbs*. This was an underpinning of the position he took in reviewing a number of publications in the same journal on life and organization the previous year when he announced: "A study of morphology presents us with so many facts inexplicable on the ordinary teleological theory, that at the present day there are few comparative anatomists who have not given in their allegiance to evolutionism."[142]

In 1872 Macalister contributed another article to this publication, now called the *Dublin Journal of Medical Science*, in which he explicitly endorsed Owen's hypothesis of "a fundamental unity of plan in animals" but claims that if such an explanation is analyzed, "we will find that though the expression annunciates an undoubted fact, that such a unity of plan exists, yet it contains no information or interpretation as to the cause of the unity, which after all is the point of which we are in search; for if muscle anomalies be the accidental production in an animal of the normal structures of another, it is the cause of this, whatever it may be . . . which must determine the production of anomalies." Macalister offered as an explanation the Darwinian theory of reversion by suggesting that anomalies are most likely due to the liability of the frame to return in some of its details to more primitive arrangements. Arrests of developments were, he said "easily explained on an evolution hypothesis."[143] Here he was playing to his Darwinian readers more explicitly, but without compromising his structuralism by an endorsement of natural selection.

This interplay of "location and locution" in scientific communication can be observed again in his 1886 publication of "Man Physiologically Considered." Here in section 6, he celebrated the body "as a monument of design, the fitting of means to ends, an evidence of the wisdom of God," and developed his piece around the teleological premise that "man stands forth as the crown of creation."[144] Again, attending to the strategies deployed in different textual arenas of the origins debate, we see how Macalister's writing is shaped. This piece was written specifically for the Religious Tract Society and befits the requirements of this evangelical publication whose objective was to provide the working classes with books that popularized science with a proper "Christian tone."[145] Each Religious Tract Society volume had to include a statement of the Atonement, which almost always appeared at the end of the book so as not to obstruct the flow of the narrative. It is in this context of the reconciliation of "man" to God that Macalister's concluding section must be read, stating "that the prominent lesson which the examination of the human body impresses on us is that of perfect adaptation of means to ends, of structure to function. The unprejudiced mind cannot fail to read in every organ, nay, we may say in every cell and fibre the inscription of purpose, and to learn thereby that they are the products of supreme power directed by supreme wisdom."[146] However, despite the distinct overtones of utilitarian natural theology, the piece has no argument with Owen's earlier statement: "To what natural laws or secondary causes the orderly succession and progression of such organic phenomena may have been committed, we are as yet ignorant." So yet again Macalister maintains his structuralist credentials, but this time provides his pronouncement with a more overtly Paleyan gloss.

Because Owen's archetype reinterpreted as a Platonic idea could be integrated into a design argument, Rupke notes that there was "a strategic retreat to the safe grounds of Paleyan function in some scientific quarters in the mid to late nineteenth century." Richard Yeo has argued that this Platonization of Owen's archetype by primarily Oxbridge Paleyites represented an attempt to affirm the assumptions of natural theology.[147] Macalister explored this idea of the Platonic archetype—where it was construed as the "fullest, highest and most complex of types"[148]—in his appraisal of evolution in church history. Here, he referred to the original form of the church as the "perfect and complete institution," but then, by comparing it to the "opposing thesis," that is, a church that "while organized on certain broad general principles of spirituality, unity, and universality, has in it an element of growth and expansion correlated to the mutations of the conditions with which it is surrounded," he called on readers to "endeavour by a process of induction, to determine whether the history of these portions of the church system favours the evolutional or archetypal theory."[149] We see here the innovative ways in which Macalister explored idealist notions of development, by theorizing in a theological rather than a scientific textual space.

Finally, some reflections on the anatomy laboratory are in order. Of Macalister's disciplinary spaces in the 1870s and 1880s, those encompassing pedagogy predominated. As we have observed, his powers of memory and persistent accumulation of facts by personal observation resulted in a detailed knowledge of anatomy and the literature relating to it that was uncanny, and at times disconcerting to students and those who sought his advice. Elizabeth Hurren gives us an insight into the way in which he redesigned the system of "practical teaching" at Cambridge to great effect. Under his direction anatomical studies became more regimented, regulated, and supervised, highlighting Steven Shapin's suggestion that domination, drilling, and discipline are the keys to understanding how scientific knowledge spreads from local venues to global circulation.[150] Bodies were to be dismembered to create a wider range of "teaching materials." For example, he estimated that 188 students would require 56 dissecting hands, 40 arms, 32 legs, 32 abdomens, and 8 thoraxes.[151] In the teaching environment of the anatomy lab, Macalister's cadavers in Dublin and Cambridge became "subjects," "tissues," "projects," "samples," or "studies" to be used in "medical advancement," "for the benefit of humanity." All these terms reflect the infringement on Macalister's "transcendentalist" terrain by the pedagogic spaces of knowledge promulgation, in an era when biologically focused departments at Trinity and Cambridge were asserting their credentials as modern centers of science. These darker fieldwork spaces—where the medical profession was complicit in activities that were construed in some quarters to be detrimental to the soul—had to be carefully negotiated by Macalister. He rose to this challenge, as we see, by an emphatic endorsement of his fieldwork activities as "disembodied" medical practices. So, by probing the spaces of empirical precision, he found a pedagogic focus around which he could make pertinent observations relating to wider debates, without compromising his scientific or religious credibility.

In keeping with recent spatial inquiries that construe late nineteenth-century encounters between science and religion as products of particular geographical locations, this chapter explored the life-spaces of the Presbyterian biologists David Moore and Alexander Macalister. Both men were resident in Dublin at the height of the origins debates following Darwin's publication of *On the Origin of Species*. Here attention was drawn to the role of form and function in this debate and the considerable extent to which the primary tenets of morphological science, structure and function, coacted with the late Victorian city spaces of these scientists.

The themes of morphology and local space were interwoven throughout, on the one hand, elucidating how scientific precepts of form were made to work against Darwin's theory in different ways depending on the specifics of place. And, on the other hand, opening up nineteenth-century evolution debates to the scrutiny of life-space, revealing further ways in which geography can unsettle traditional readings of the science and religion encounter. For David Moore, the practicing of science at home perceptibly changed both the home and the science, and beyond familial concerns, the gardens acted as a link between his convictions as a practicing Presbyterian and his public role, bringing natural theology and the purposeful workings of a beneficent designer to his visitors. In addition, in the political climate of late nineteenth-century Dublin, the translation and negotiation of classification systems propelled his

FIGURE 9. Trinity College Dublin ca. 1900.This image is reproduced courtesy of the National Library of Ireland, NLI, Limp, 2513.

botanical work into the national arena. The notion of uplift became a unifying theme in Moore's life, blurring boundaries between the domestic and the scientific, individual and community, body and soul, and life and death, all the while underpinning a Paleyan natural theology suggestive of an earlier era, but existing "in harmony" and "contentedly" with Moore's late Victorian Dublin city spaces.

By the 1870s, the functionalist worldview changed paradigmatically and the traditional utilitarian unity of Paley's world became in the minds of a new generation of scientists a materialistic unification of mind and matter. Morphologically speaking, Macalister's will to keep these issues distinct and separate found refuge in structuralist biology. His cityscape—from the dark spaces of anatomical acquisition to the light spaces of knowledge promulgation—also promoted a bounding rather than a blurring of spatial arenas. This discrete and dualistic theme turned him away from Darwinian spaces to find alternative mechanisms that he hoped might be made to accommodate both physical and mental evolution. His belief that the religious and moral capacities of the psyche were beyond the scope of an evolutionary framework and that the "soul" was a unique human attribute set him on a collision course with progressive evolutionists. But in addition to this, unlike Moore, Macalister had to *search* for space to express his theism, and when science was found wanting he explored other academic avenues. Finally, by embracing a suite of late Victorian Dublin rhetorical spaces, he effectively negotiated the diverse evolutionary stances of these public and private personae.

Chapter 3

Human

In June 1898, at a Dublin meeting of the Anatomical Society of Great Britain and Ireland, delegates were invited to the Trinity Anthropometric Laboratory, where they "submitted their heads to the anthropometric test." Commenting on the cephalic indexes gathered—which ranged from "extreme brachycephaly to ultra dolichocephaly"—the team concluded, "It is clear therefore that there is no special form of head required to *make an anatomist*."[1] This moment of *ex camera* self-scrutiny at the Dublin meeting serves to introduce the topic of chapter 3—Victorian predilections around the study of humankind to define and construct "types."

The anthropometric venture during the last decade of the nineteenth century occupies a shady place in the history of Irish science, and its legacy has only recently begun to be explored.[2] The comprehensive archive of Island and coastal inhabitants compiled by the peripatetic laboratory in the west of Ireland has for a long time been the primary source available for scholars to ponder on the juxtaposition of "imperial" anthropology and a "primitive" Irish populace.[3] As a result, the Western Seaboard, particularly the Aran Islands, has been in the spotlight as a fitting arena in which to examine the conflict of cultures engendered by Victorian science. In addition, a recurring theme across the spectrum of scholarship on the Anthropometric Laboratory has been a focus, or at the very least, substantial allusion to the wide-reaching influence of the 1859 publication of *On the Origin of Species* and the ensuing "Darwinian" nature of colonial social policies.[4]

In something of a departure from both these preoccupations, in this chapter I would like to move the site of study back to Dublin, draw the intellec-

Cephalic Index.

W. Spalteholz,	86·5	H. Gadow,	79·0
C. R. Browne,	85·9	J. Disse,	78·9
*J. Kollmann,	85·2	G. Elliot Smith,	78·6
G. B. Howes,	84·7	*D. Hepburn,	78·5
Hector Leboucq,	83·9	A. H. Young,	78·4
F. Frohse,	83·5	Stanley Boyd,	77·9
E. Barclay Smith,	83·3	Sir William Turner,	77·8
E. H. Taylor,	82·5	R. J. Berry,	77·6
J. Cantlie,	82·4	A. Keith,	77·3
B. Windle,	82·2	D. J. Cunningham,	76·8
C. de Bruyne,	81·6	A. M. Paterson,	76·4
A. Birmingham,	81·5	——	74·6
A. F. Dixon,	80·9	R. Howden,	74·4
*J. Symington,	80·7	G. D. Thane,	74·2
James Musgrove,	80·4	J. Lynn Thomas,	73·8
Wilhelm His,	79·7	A. Macalister,	73·0
C. D. Marshall,	79·4	T. H. Bryce,	72·7
A. Robinson,	79·3	P. Thompson,	72·2

FIGURE 10. Cephalic Indices of Prominent Anatomists, Dublin 1898. From Cunningham and Browne, "Report from the Anthropometric Laboratory."

tual focus somewhat away from Darwin, and explore instead the extent of pre- and non-Darwinian ideas of human origin apparent in the aims and methodologies of the Trinity laboratory. Taking this approach will position the workings of the laboratory in the context of a wider nineteenth-century anthropometric episode, with which Dublin city spaces briefly interacted in the 1890s. The chapter then turns to examine the spaces of Alfred Cort Haddon (1855–1940) and Daniel John Cunningham (1850–1909). These scientists were placed at the helm of the Dublin site at the height of the Home Rule debate, and were charged with implementing its progressive methodologies, but both men struggled variously with its stated anthropometric goals due to personal scientific motivations. Although chapter 3 continues to engage with the life-space theme of previous chapters, it employs a slightly different format in order to encompass the spaces of the Anthropometric Laboratory at Trinity as an important site around which the lives of both scientists revolved between 1891 and 1900.

The establishment of the Anthropometric Laboratory in the Trinity College Anatomy Department in 1891 was the Dublin manifestation of an intellectual storm that swept biological science into the realms of politics and brewed throughout the nineteenth century on both sides of the Atlantic. At the eye of the storm was the issue of racial origin. The question of whether racial variations were specially created or they developed over time was pivotal to nineteenth-century anthropological discourse. Before Darwin, the inquiry centered around whether there was one Adam and Eve or multiple creations. The monogenists argued for one origin of races, and the polygenists argued for several distinct origins.[5] After Darwin the discourse shifted slightly to inquire into whether the species had one origin with several evolved racial variations or each race constituted its own species having only very remote connections to others. But from a broader point of view, George Stocking discusses how polygenism and monogenism can be regarded as specific expressions of enduring alternative attitudes to human variety in the post-Darwinian era. The durability of polygenist thought in particular, he claims, "provides an interesting example of the way in which certain ideas may persist—modified in various ways, juxtaposed in new combinations, but at bottom, relatively unchanged—long after the controversies in which they were originally advanced."[6]

It is this ideology and its embodiment in the anthropometric venture that informs the first part of the story. The focus then moves to the site of anthropometric inquiry in Dublin where the new science was embroiled in two important and interconnected nineteenth-century debates: the question of Irish origins, and whether or not Home Rule should be granted to Ireland. Here the aim is twofold. First, I want to situate the broader scientific venture within this local political landscape. Second, I want to examine how the personal and political life-spaces of its two key players influenced and were influenced by the new science of anthropometry in Dublin.

By the mid-nineteenth century in the United States, Thomas Jefferson's rallying call to a newly aligned populace, that "all men are created equal" and had certain "unalienable Rights" to "Life, Liberty and the pursuit of Happiness,"[7] began to ring hollow, as Enlightenment natural history attempted to answer new questions about the morality of slavery and prejudice raised by the American Revolution. Despite calls by African American intellectuals to secure equality in the postwar period, new systematic theories of race began to emerge.[8]

Some of these had their origins in the neotraditionalist school of polygeny, which held that humans and animals were created essentially where they were found and that God created different races to fit distinctly different environments. According to the philosopher of science Stephen Asma, polygenist thinking both before and after Darwin was oriented by the internal metaphor of the human self. Asma explains that according to the polygenist view, the African descendant, the Native American descendant, and the Caucasian descendant, all represented fixed kinds of diverse entities and took their diversity from some innately bestowed respective essence.[9] Asma explains that the concept was derived from Aristotle's vision of the natural slave: "For that some should rule and others be ruled is a thing not only necessary but expedient; from the hour of their birth some are marked out for subjection."[10] In this view an internal soul immune to environmental conditions was said to eternally define the slave from ancient times to the present. This conceptual placing of races was easily transposed to the late nineteenth-century promulgation of racial hierarchies.

By the end of the century, anthropologists had joined the ranks of zoologists and geologists who had grown accustomed to the idea that their science was essential to improving the management of state resources, ranging from colonized peoples to edible fish,[11] and in this environment the "internal soul" of the neotraditionalist morphed into the "biological entity" of secularized Anglo-American polygeny in the late Victorian period. Post-Darwinian polygeny therefore found refuge in an unholy alliance of evolution and numbers—a scientific racism that had its roots in separateness and difference. As evolution became the guiding principle for life and the human sciences, imbuing many older ideas with a scientific respectability, the pre-Darwinian idea of stable racial types gained new force and acquired new evolutionary justification on both sides of the Atlantic. The model that gained ascendancy was a potent blend of Darwin's emphasis on struggle and Spencer's emphasis on progress, conflating their versions of evolutionary theory into a single worldview. Essentially, this resulted in the reading of Darwin through a Spencerian lens as "social Darwinism." In this worldview, evolution occurred in society as it had in nature. Society could achieve evolutionary progress only by a fierce struggle for survival, with the losers ending up at the bottom. Just as newly scientific physical anthropology helped to legitimize older concepts of racial type, social Darwinism lent new respectability to well-worn notions of racial hierarchy and interracial struggle.

In the United States, increasingly negative characterizations of the black and Native American populations as inferior products of nature were constructed. Writing in 1852, the African American physician and writer James McCune Smith's use of the term "negro" was in reference to intellectually constructed blackness. It denoted "not an actual physical being of flesh and bones and blood, but a hideous monster of the mind."[12] These polygenetic ideas on racial construction traversed the Atlantic on various political vessels, eventually to alight on the foundational fabric of Dublin's Anthropometric Laboratory. Their post-Darwinian fate here is examined by means of the institutional, disciplinary, and political spaces of Daniel John Cunningham and Alfred Cort Haddon, to elucidate a very local negotiation of alternative anthropological meanings behind the imported polygenetic facade of the new laboratory.

Mid-nineteenth-century British polygeny centered on the Anthroplogical Society of London, James Hunt, and the Knoxian-inspired belief that archetypal plans, not external environment, best explained organic diversity.[13] From here it was a small step to the pronouncement that different races were separate and gradable biological entities with profound physical and moral chasms between them. This concept of difference became ingrained in anthropological thought even as polygeny itself lost favor from the 1860s onward. For some, the "scientific anthropology" represented by the new Dublin Anthropometric Laboratory, fashioned on Francis Galton's London endeavors, provided precisely the kind of data that could be applied to these graduated racial hierarchies, and that in turn suited the needs of imperial economics on this side of the Atlantic. Thomas Carlyle, who coined the term "the dismal science" for the classical economics proposed by John Stuart Mill and Adam Smith, preferred the notion that some races were inherently lazy and would fail to lead productive lives unless forced to do so.[14]

The context for this nineteenth-century debate between classical political economy and its critics has been located by Gordon Bigelow in the Irish question. Although Darwin's theory of natural selection almost foundered on the Irish question in the 1860s, the mid-nineteenth-century leaders of the Anthropological Society of London devoted a great deal of attention to the problem of whether the Irish constituted a separate race. Against these considerations were the changing attitudes to colonialism engendered by the Jamaica revolt (Morant Bay rebellion) in 1865, which had polarized opinion in England regarding imperial matters.[15] Supporters of John Eyre's harsh put-down of the rebellion, such as Charles Kingsley, drew parallels with Ireland and declared that the realities of life demonstrated how "differences of race are so great that certain races, for example, the Irish Celts, seem unfit for self-government." Carlyle and others believed that parliamentary reform would "unleash at home those primitive horrors displayed in Jamaica."[16] W. R. Greg, in an attack on John Stuart Mill—who argued that land reform would help solve the problem of poverty in Ireland—summed up the Carlylean point of view, "Mr Mill never deigns to consider that an Irishman is an Irishman, and not an average human being . . . an idiomatic and idiosyncratic, not an abstract, man."[17]

However, for others, Britain's policy toward Ireland differed from that realized in other colonies because it admitted the possibility of full assimilation. In this view, the fundamental question for Ireland's future was whether its inhabitants were capable of assimilating English culture and behaving responsibly as full citizens of a parliamentary regime. Debates over Irish assimilation focused on two areas, land reform and Home Rule, and in both debates anthropological arguments figured in the decisions. The view that the Irish could be assimilated—which was being backed up by ethnological surveys pointing to the "mixed stock" of the British Isles—was expressed by prominent members of the anthropological community such as John Lubbock and Thomas Huxley, and this justified their opposition to Irish Home Rule. According to Henrika Kuklick, when anthropologists opposed Irish Home Rule on the grounds that the Irish were not racially distinguishable from other British citizens, they were in fact articulating a clearly liberal political position.[18] And many opposed Home Rule with arguments based on the work of authors such as Ernest Renan, in which even seemingly complimentary generalizations about the Irish national character could, in the Victorian context, be counterproductive for Home Rule.[19] The attribution of qualities such as "poetic, light-hearted, imaginative, emotional, playful, passionate, and sentimental" meant that strong outside guidance and governance were essential.

Financially, it was feared that the very prospect of a Home Rule bill would cause a serious withdrawal of capital from Ireland. Carlyle foresaw this when he commented: "To at least render Ireland habitable for capitalists, if not for heroes; to invite capital and industrial governors and guidance (from Lancashire, from Scotland, from the moon, and from the ring of Saturn); what other salvation can one see for Ireland?" Anti–Home Rule arguments asserted that Nationalists did not have the confidence of capitalists and cited the falling price of Bank of Ireland stock and other Irish securities on the announcement of Gladstone's Bill.[20] So, the imperial progressive stance invested much in reductionist philosophies drawn from biology to justify free market economics, and into this mix filtered the methodologies, measurements, and instrumentation newly acquired by American statisticians following the Civil War. A classificatory surge ensued, and buoyed by an unquestioned faith in numbers, assigned roles and levels to every aspect of humanity.

Such was the intellectual context for the emergence of the Dublin Anthropometric Laboratory. The laboratory, the first of its kind in Ireland, was a welcome addition to the growing scientific armory of the metropole. Here, however, the new venture at the culturally dominant and prestigious institution of Trinity College is not treated as though it were an entirely autonomous scientific center of action. Instead, the Dublin lab is viewed as an interstitial entity drawing its value and meaning, as Robert Kohler suggests, from the people and institutions with which it was juxtaposed, opposed, and integrally networked.[21]

The Dublin site can be scrutinized on many theoretical planes. It was a "staging post" in the formalization of ideas on racial origins and order, which had emerged with the U.S Sanitary Commission and its military anthropometric stations. Alternatively on this side of the Atlantic, it could be seen as a site of justification for those who wished to transpose "scientific facts" to a political key and bring an anthropological dimension to the Home Rule question. On yet another plane it could be examined as an offspring of Francis Galton's investigatory framework for biometrical work on inheritance, this aspect gathering steam at the turn of the century when Daniel John Cunningham became involved in survey work for the Privy Council Committee on physical deterioration.[22] But whatever plane is chosen, the scent of antebellum polygeny was never far from its empirical pretentions.

By placing the laboratory within these wider ideological spaces emanating from centuries of global exploration, the American Civil War, British imperial economics, and evolutionary biology, it becomes clear that procedures devised for earlier polygenetic theories of multiple racial origins were inherent in the methods and instrumentation of the "new science" of anthropometry in Dublin. At the core of this methodology was an unerring belief in the efficacy and power of numbers, and numerical authority was marshaled to differentiate between and separate racial groups. An understanding of the way this close relationship between polygeny and anthropometry developed is key to acknowledging the polygenetic legacy of the Dublin laboratory.

To begin with, physical anthropology in the nineteenth century was dedicated to one major aim: the measurement of human bodies, particularly heads, in order to identify the stable racial types underlying human populations. The basic assumption of the science was that each race could be represented by its own essential set of traits, which were in turn associated with corresponding mental and moral characteristics. As the discipline developed, anthropologists believed they could extract and identify the essence of a race by techniques of careful measurement.[23] A nascent anthropological science thus absorbed earlier nineteenth-century attempts to classify human beings into types, arrange them in hierarchical order, and imbue the gradations with distinct peculiarities. The imprint of polygeny gained depth as the process intensified. Johann Friedrich Blumenbach had designated five races or varieties of man in the second edition of his treatise *On the Natural Varieties of Mankind* (1781), and his classification system underpinned the importance of Caucasian as the basis from which the other races derived.[24] These ideas then became subsumed into a rough hierarchy of human races that eventually became part of accepted nineteenth-century anthropometric wisdom.[25] Although many of the early classifiers were not advocates of separate racial origins, their systems were later adopted and fixed by polygenist thinking. Stephen Jay Gould's contention that "anthropometrical methods became a search for characters that would display the correct ranking of a presumed order and not a numerical exercise in raw empiricism" was aptly displayed in the methodological assumptions of the Dublin Anthropometric Laboratory where measurements were gathered of "students of Trinity College Dublin" and of "peasants of the farming and fishing classes in the West of Ireland."[26]

The team at the Dublin Anthropometric Laboratory, by collecting, classifying, and interpreting numerical information, would have regarded themselves in the words of Gould as "servants of their numbers" and "apostles of objectivity."[27] But Bruno Latour's observation that "knowledge does not reflect a real external world that it resembles via memesis, but rather a real interior world, the coherence and continuity of which it helps to ensure" is quite pertinent here.[28] An analysis of the rigorous methodology and systematic measurement in the Dublin venture does indeed reveal a system that carried within itself a framework of assumption rooted in the polygenist tradition. This system was designed to differentiate between species and races and could be marshaled for political means, and if necessary, demonstrate that the Irish were a separate species.

The Dublin Anthropometric Laboratory

On Saturday June 13, 1891, Dr. Samuel Haughton, a professor of geology at Trinity College, obtained leave to issue invitations in the name of the provost and fellows to members of the university and others to attend the opening of the Dublin Anthropometric Laboratory on June 25, at the Trinity Medical School.[29] The Dublin University calendar for that year notes that "a portion of the Zoological museum has been fitted up for an anthropometrical laboratory, open on Tuesdays, Thursdays and Saturdays from 2pm to 4pm, free of charge."[30] The laboratory commenced operation under the direction of Alfred Cort Haddon, a professor of zoology at Royal College of Science for Ireland, Daniel John Cunningham, a professor of anatomy at Trinity College, and Charles R. Browne, Cunningham's assistant. Since anthropology as a discipline was not then taught at Trinity, Browne demonstrated anthropological methods for two hours, three days a week to interested students.[31] Instruments were provided by a grant from the Royal Irish Academy each year, and purchased by a Royal Irish Academy committee composed of Haughton, Cunningham, and Haddon. Every year the equipment was taken to selected districts in Ireland for a systematic study of the inhabitants.[32] A letter from E. P. Wright to Haddon on July 10, 1893, includes a check for £20 "to assist in prosecuting ethnological in-

FIGURE 11. D. J. Cunningham at work in the new Dublin Anthropometric Laboratory. Reproduced by kind permission of the Board of Trinity College Dublin. TCD MS 10961/5 41.

vestigations in the remoter parts of Ireland . . . then to be understood that any material collected is at the disposal of the Academy's Committee."[33] A glimpse into this late Victorian Dublin laboratory and how it connected with global spaces can be afforded by one of the university's tercentenary publications: "On the ground floor is the Anthropometric Laboratory where measurements and records are taken on a somewhat more extended plan than that introduced by Francis Galton at South Kensington and a metric system of notation has been adopted similar to that in use on the continent of Europe, especially in Paris and lately introduced into the Anthropometric Department of the Military Medical School of Washington."[34] The explicit adoption of Galton's measuring schemes in Dublin at once locates the laboratory's debt to a racial science informed by polygeny.[35] By Galton's own admission, "The *cui bono* of making measurements is to define the individual or the race and to show in what way, and to what extent he or it differs from others," and the metric system installed in Dublin was only one facet of the notational armory adopted from Paris and Washington to give scientific validity to such ideas.[36]

Aided by the precision of physics and the new statistics, this "numerical authority" called on by the Dublin laboratory had been securely harnessed by Civil War anthropometry through the involvement of Joseph Henry and Adolphe Quetelet. Henry, the first secretary of the Smithsonian, was a physicist with a background in electricity, magnetism, and meteorology, and under his direction the new Smithsonian Institution was "piloted away from politics, commercialism, and speculation." He felt that in the human sciences speculation and policy too often took precedence over empirical research. As editor of the first volume of the Smithsonian Institution's *Contributions to Knowledge* series he demanded rigorous scientific standards and used the Smithsonian's financial resources and status to redefine the proper methodology for publications on anthropology. The series from the Smithsonian set a new standard in anthropology and led out of the speculative period into what American historians Gordon Willey and Jerome Sabloff have called the "classificatory-descriptive" period.[37] Although Henry's scientific contributions to anthropometry were amenable to a polygenetic mindset he was personally at odds with the secular polygeny of his fellow American George Gliddon, whom he accused of having misrepresented scientific facts to attack religion.[38]

As Stocking notes, post-Darwinian anthropological discourse became more free-floating, and scientists who were explicitly monogenetic might hold quite polygenist ideas on a particular racial issue, for example, miscegenation.[39] As a result one can find numerous manifestations of what one could call polygenist thinking emerging from otherwise monogenetic quarters toward the end of the century. Also on a practical level, the Sanitary Commission in Washington used the design of Joseph Henry's anthropological measuring devices, along with those of consummate polygenist thinkers, in order to facilitate uniformity in instruments and procedure and he thus became implicated with the polygenetic venture.[40]

The new statistical methods of the Belgian mathematician Adolphe Quetelet were also called on by the United States Sanitary Commission in the attempt to create an "average man" as representative of specific groups within the army, and analyses were undertaken of that specimen in his various relations—physical, social, and moral. Quetelet was one of the first to apply statistics and probability to the social sciences. In his book *Sur l'homme et le développement de ses facultés, ou essai de physique sociale*, he outlined his concept of a social physics and describes his "average man" as being characterized by the mean values of measured variables that follow a normal distribution. The statistical methods of Quetelet were enthusiastically pursued by the Dublin laboratory as they tabulated hair and eye color, nigrescence, and both cranial and postcranial measurements to ascertain an average type for each island.[41]

A lab recording sheet dating to September 28, 1892, for a Michael Mullen of Aranmore, contains an annotation in the margin that highlights Haddon's contention that this man's measurements place him (like Quetelet's concept of "l'homme moyen"), as a typical man in all respects for that island.

The figure of Adolphe Quetelet was instrumental in the application of hard science methodology to anthropology and stands out as a foundational influence on the methodology that was implemented in Dublin. Drawing on the statistical methodology of Quetelet, the Dublin lab continued the process that had brought meaning to vast amounts of data accumulated during the American Civil War.

The presence of such vast quantities of skulls and the marked bias toward cranial measurements on the anthropometric forms also underpin the importance of head measurements to the new venture. Craniometry had been introduced to anthropometry by John Beddoe's French colleague and friend Paul Broca. It was an attempt to advance the practice of phrenology and bring more scientific authority to the study of heads. Broca invented over forty instruments for the more accurate study of skulls and, along with the application of the cephalic index, anthropologists could now differentiate between dolichocephalic, brachycephalic, and mesocephalic head forms. According to Gould, Broca's cardinal bias lay in his assumption that human races could be ranked in a linear scale of mental worth and his methods "determined the relative position of races in the human series."[42]

Broca's ideas were taken up enthusiastically by Samuel Morton, who attempted to put the Frenchman's assumptions to the test. Morton, seen as the great data gatherer and objectivist of American science, the man who raised the immature enterprise of phrenology from "the mines of fanciful speculation," held Broca's hypothesis that a ranking of races could be established objectively by physical characteristics of the brain, particularly by its size. He embarked on an extensive skull-collecting mission to test the theory and defended the status of human races as separate created species in *Crania Americana.*[43]

An ossuary resembling Morton's "American Golgotha," albeit on a smaller scale, was re-created in the Dublin laboratory.[44] Hundreds of craniums from different districts in Ireland were gathered and stored along with skulls and skeletal material from the collection of William Wilde. Also lining the walls were cabinets containing craniums from the phrenological collection of John Grattan of Medical Hall, 12 Cornmarket, Belfast.[45] Osseous material from archaeological investigations was also deposited with the laboratory for analysis, some of which was housed permanently, and other samples such as those from Colonel Plunkett, the director of the Museum of Science and Art Kildare Street, were deposited on loan to the Trinity College Museum.[46]

FIGURE 12. The Skull Cabinets, Dublin Anthropometric Laboratory. Reproduced by kind permission of the Board of Trinity College Dublin. TCD MS 10961/5 41.

One problem previously encountered with the head-height measurement, which involved three rather than two reference points, was the reliability of the third point.

In 1861 George Busk came up with a craniometer that measured two planes at right angles. But to further ensure their reliability, Cunningham adapted and refined Busk's craniometer for the Dublin measurements.[47] The head-height measurement was an important component in ascertaining the cephalic index and thus differentiating between dolichophalic, mesocephalic, and brachycephalic types, but more pertinent to those with polygenetic tendencies was the inherent intent of the new enterprise to measure facial angle. Although Charles R. Browne does not list a facial goniometer, an instrument that measures facial angle directly, among the lab apparatus, we do know that the Dublin Anthropometric Laboratory was measuring for prognathism in the 1890s.[48] The English polygenist and anthropologist Sir Arthur Keith, who championed the notion that racial conflict was a major factor in human progress in *The Antiquity of Man*, discussed "the facial triangle and the method of estimating the degree of prognathism by comparing the lower side of the triangle, the auriculo-alveolar distance, with the upper side of the triangle, the auriculo-nasal distance."[49] It is evident from the Dublin record sheets, that measurements of auriculo-alveolar length and auriculo-nasal length were required for its anthropometric project.

The determination of prognathism, or measurement of the "facial angle," was restored to science by Petrus Camper in 1784. This ancient Greek formulation became the most extensively used criterion for determining racial difference in the nineteenth century. It gave preeminence to the Caucasian by virtue of having its origin as an artistic ideal that gave substance to the statuary scale of beauty in Greek art. Although Miriam Claude Meijer has demonstrated that Camper was essentially a monogenist who used the idea of the facial angle to show the relationship between African and European specimens and thus to promote equality, others such as the anatomist Charles White invoked facial angle in support of polygenism. In his *Account of the Regular Gradation in Man*, White states, "In whatever respect the African differs from the European, the particularity brings him nearer to the ape."[50]

As well as measuring for facial angle, the template form included in Haddon's address to the Royal Irish Academy in June 1892 contained forearm measurements and the measurement of the tip of the mid-finger to the patella. These measurements were part of a series of modifications made to forms and apparatus by the US Sanitary Commission, aided by the polygenist Louis Agassiz, to include black and Native American soldiers.[51] After the Civil War, in an effort to continue uniformity in anthropometrics, this modified material was distributed among colleges and institutions for continued research, and by the time it arrived in Dublin it was an intrinsic part of the anthropometric enterprise.

At the local level, however, this racial science was negotiated in a more nuanced way. In September 1892, the first Anthropometric Survey began on the largest of the Aran Islands when Haddon and Browne set up a temporary recording station in front of a cottage on Aranmore and measured the islanders. The data were registered on forms (see figure 13) and it soon became apparent that certain measurements had not been taken by the Dublin group. Shortly after the work began, the team made clear the doubts they had about the mid-finger to patella measurement. The following extract is from a paper read to the Royal Irish Academy by Charles R. Browne on December 4, 1891: "Very soon after the regular work of the Anthropometric Laboratory of Trinity College was started, it became evident that there was need for some slight improvement both in methods and instruments. The first change was the abolition of the measurement from tip of mid finger to centre of patella as this was found to be difficult, I may say impossible to get correctly owing to the very great discrepancies made by slight differences in the attitude of persons being measured." The practical difficulties outlined do not appear to

THE DUBLIN ANTHROPOMETRIC LABORATORY.

The Laboratory is in the Museum of Comparative Anatomy, Trinity College, Dublin. The Laboratory is open to the Public from 2 to 4 p.m., on Tuesdays, Thursdays, and Saturdays.

Date of Measurement.			Initials.	Sex.	Birthday.			Eye Colour.	Hair Colour.	Page of Register.
Day.	Month.	Year.			Day.	Month.	Year.			

Head Measurements.							Stature.		Limb Measurements.					
Cranial length.	Cranial breadth.	Cranial height.	Auriculo-alveolar length.	Auriculo-nasal length.	Face length.	Face breadth.	Height standing, less heels of boots.	Height sitting above seat of chair.	Tip of midfinger to styloid.	Tip of midfinger from centre of patella.	Styloid to Epicondyle.	Epicondyle to Acromion.	Acromion from ground, less heels of boots.	Span of arms.

Weight in ordinary clothing (in lbs.)	Strength of Grasp (in lbs.)		Greatest speed of blow in feet per second.	Breathing capacity (in Cubic inches).	Colour sense.	Keenness of Eyesight. Distance in mm. of reading standard numerals.		Is the Sense of Hearing normal?	Highest audible note by whistle (Vibrations per second).	Reaction time, in hundredths of a second.		Left Thumb.	Right Thumb.
	Right hand.	Left hand.				Right eye.	Left eye.			To sight.	To sound.		
									, 000				

One page of the Register is assigned to each person, in which his measurements at successive periods are entered in successive lines. A copy of these made at any specified date may be obtained on application by the person measured, or by his or her representative on receipt of a stamped envelope.

N B.—In the case of students, it is very desirable that their measurements should be recorded every six months, so that they may thereby have an accurate means of estimating their physical as well as their mental development.

The Anthropometric Laboratory of Ireland. 39

FIGURE 13. Anthropometric Data Sheet from the Trinity Laboratory. From Cunningham and Haddon, "Anthropometric Laboratory of Ireland."

have been highlighted in the American reports, but presenting the problem in this way allowed the Dublin team to avoid both political and scientific debate about the methodology at such an early stage in the academy's important new scientific venture.

The polygenetic presumptions and implicit scientific racism of the new lab's methodology did not sit well with all of Haddon and Cunningham's spaces in Trinity. Take, for instance, the discarded measurement outlined by Browne above, which applied to the distance from the finger tip to the patella—this measurement was essential for assessing upright posture. According to the procedures of the US Sanitary Commission, the difference in the distance here between the black and white soldier marked the black soldier out as that much closer to the anthropoid in development.[52] It was considered by the commission that this difference was due to the greater arm length and shorter body length of the black physique. Although the use of measurements to arrange races in this hierarchical manner, with Caucasian at the top, could equally benefit the monogenist viewpoint, the commission's scholarly emphasis appeared to be biased toward unearthing essential racial differences that ultimately benefited the polygenetic rationale for slavery and gave a semblance of scientific authority to any social or political objective that mooted racial segregation.

In leafing through late Victorian anthropology journals, it becomes clear that the discussion around posture, which cropped up intermittently throughout the 1890s, was infused with the preconceived notion that the primary role of measurements was to differentiate between races, rather than to highlight evolutionary links and racial homogeneity. Arthur Thompson's examination of "the influence of posture on the articular surfaces of the tibia and astragalus in the different races of men and the higher apes" in the *Journal of Anatomy and Physiology*, and D. K. Shute's exploration of racial anatomical peculiarities in the *American Anthropologist*, in which he unearthed "an interesting series of correlated changes in the thorax, pelvis and lumbar vertebrae, with the assumption of the upright posture," are indicative of this tendency.[53] To compound the issue, although presented in seemingly innocuous tones, comments about the simian nature of the African pelvis, vertebral column, or limb form clearly made connections between so-called inferior races and the ape.[54] Whether marshaled for polygenist or monogenist arguments of racial order, these measurements had their genesis in a polygenetic viewpoint, which advanced separate racial origins and located constructions of inferiority on a gridded page. Some of the measurements on the anthropometric field sheets printed for the Dublin laboratory were, however, to present problems for Haddon's team.

The fact that Cunningham had swum against the polygenetic tide with his theories on posture is significant here. The mid-finger to patella measurement would have been rendered superfluous by his research on the lumbar curve in man and apes.[55] As he explained to the academy in 1886, there had been some erroneous conclusions from macerated skeletons. Because of the neglect of intervertebral discs in the constitution of the curves, he "enforced the necessity of considering the close correlation which exists between structure and function in the vertebral columns" and, according to his obituarist, was consequently able to guard against the error of supposing that "the osseous structure of the column in the lower races was a sign of inferiority: whereas, in fact it was only proof of the greater range of mobility."[56] This example highlights one of the ways in which the methodologies of the polygenetic machine were negotiated at the local level. By paying closer attention to the institutional and disciplinary spaces of the team, the interpretive milieu in which polygeny was encountered can now be exposed, and its diverse interpretations explored. Before their work commenced at the new anthropometric lab both men were immersed in the structures and strategies of other Dublin institutions, aspects of which they brought with them to the new posts. Examining these institutions in more detail deepens our understanding of why they held certain convictions and took particular stances in their new location.

The teething pains of an aspirant Royal College of Science in Ireland, where Haddon became a professor of zoology in 1880, and the established mores of Trinity College Dublin, where he steered the new anthropometric lab from 1891 to 1901, would weave a lasting influence on the fabric of his anthropological thesis. Similarly, Cunningham's engagement with anthropometry was scored indelibly by the intellectual ethos of Trinity's Anatomy Department. The particular influence of these two sites on the development of Haddon and Cunningham's anthropological thinking therefore deserves some further investigation here

The Royal College of Science began as a vision of the future of scientific education by John Tyndall and Thomas Huxley, and was launched in 1867 as the first comprehensive attempt made in the United Kingdom by the British government to give state-aided higher scientific and technical instruction.[57] In the spatial context of Dublin scientific culture it was significant for three reasons. First, as William Barrett at the school of physics had observed: "In the College of Science, Ireland possesses an institution which, in the number of its professorships and general course of study, is more complete as *a pure school of science* than anything of the kind existing in Scotland or England."[58] Recourse to philosophy, theology, or any of the other liberal arts was not a part of the educational fabric. The college's pragmatic leanings had also attracted to its site a number of prominent Darwinian scientific naturalists.[59]

Second, the college at this time was under the control of the South Kensington Department of Science and Art, and the Royal College of Science for Ireland was seen as a training ground for employment opportunities within the empire. As one of several young English candidates who had acquired chairs at the Royal College of Science, Haddon avidly maintained an intellectual network with England, often traveling to and fro on a weekly basis. Third, there were allegations that the new appointments were made to exclude Catholics from teaching posts at the college.[60] The Royal College of Science for Ireland can therefore be seen as an educational entity metaphorically sheared from the population within which it was situated.

Haddon's theoretical framework for guiding the anthropometrical investigations was set out in a series of lectures on ethnology given in 1893 at the Royal College of Science. When he promoted the use of anthropometric methods for the purpose of "giving some assistance to the anthropologist in his endeavours to unravel the untangled skein of the so called Irish race," he did so as an employee of the college with all the geographical and intellectual distance that represented in 1890s Dublin.

Little or no systematic work had yet been undertaken on the physical anthropology of Ireland, and Haddon had no compunction about explaining the particular scientific interest of the Aran Island population:

> Anyone who has travelled through the country districts must be familiar with the different types which are presented by the inhabitants. This is especially the case

> in outlying portions of the west coast and in the islands off the main land. To take one example: the fair slight men of the North Island of Aran offer a marked contrast to the dark burly men of the middle and south islands. Then again we have in Ireland certain very old colonies. These ethnical islands if we may so term them require to be very carefully studied, and will no doubt afford valuable information concerning the persistence or otherwise of racial characters.[61]

The Royal College of Science for Ireland prided itself on the fact that it offered practical laboratory courses in the physical and natural sciences when this was not a common feature of universities in Great Britain and Ireland. As we have seen in chapter 1, William Barrett actively promoted the wider intellectual benefits of teaching "measurement" physics as "a means of education" rather than a "vehicle of instruction." It seems that the numerically authoritative environs of the Royal College of Science would have been a more natural choice for Haddon to oversee an anthropometric laboratory, especially given Paul Topinard's forthright equating of polygeny with scientific progress.[62] Perhaps a late nineteenth-century anthropometric project placed here, in an atmosphere of metropolitan pure science, origins politics, and intellectual distance from the universities, would have engaged with a polygenetic scientific racism and more readily advanced theories about the "so called Irish race." However, the correspondence of the Royal College professor of chemistry, Walter Hartley, would lead to the presumption that any plans for a new venture at this location would have been curtailed by the already severe overcrowding of the building, and the dangerous condition of the laboratories where "chunks of plaster rain[ed] down."[63]

Instead, the siting of the lab in Trinity's Anatomy Department ensured a buffer zone within which the worst excesses of a polygenetic scientific racism were absorbed. From the outset, the moral integrity of the anthropometric system was questioned, in no small part, due to the ever present influence of Alexander Macalister (see chapter 2), who had advised on anatomical matters despite his move to Cambridge. Macalister's particular bone of contention, and legacy to the lab, was his dislike of craniology. In a presidential address to the British Association in 1892, he announced: "When we in our sesquipedalian jargon describe an Australian skull as microcephalic, phaenozygous, tepeino-dolicephalic, prognathic, platyrhine, hypselopalatine, lepostaphyline, dolichuranic, chamaeprosophic and microseme we are no nearer to the formulation of any philosophic concept of the general principles which have led to the assumption of these characters by the cranium in question, and we are forced to echo the apostrophe of [Aurel] Von Torok, Vanity thy name is Craniology."[64] Macalister's successor Daniel John Cunningham inherited this dissatisfaction with the philosophical dimensions of craniometry. Cunningham privately confessed to Haddon that he hated craniology "for my whole soul revolts against it, though, I suppose it is better to keep it going till it is replaced by something better."[65] Also, the Anatomy Department in Trinity had long been concerned with research into anatomical variations. Since variation was the essence of evolution, understanding its nature was crucial to the work of the team. Macalister's engagement with alternative mechanisms of evolution espoused by structuralists, and later, the "discontinuous variations" of William Bateson, as an alternative to Darwinian gradualism, found further and more concrete expression in Cunningham's adoption of Galton's hypothesis of "organic stability" as a maxim for his research into degeneration. Both Bateson and Galton came independently to the conclusion that evolution must proceed in discontinuous steps or saltations.[66] The view of evolutionary mechanisms in Trinity's Anatomy Department was at odds therefore with the Darwinian view encountered by Haddon at the Royal College of Science. From the outset, it is clear that Haddon and Cunningham's introduction to anthropometry owed much to their institutional vantage points. How that relationship with the new science was to develop is now explored by a closer examination of the habitus of each scientist.

Haddon's Habitus

Alfred Cort Haddon, albeit engaged with the available knowledge systems and methodologies of late Victorian physical anthropology, had his eye on a longer-term goal for the discipline. Although his early lectures on anthropometry were framed by the rhetorical spaces of the Royal College, his personal aims for the future of anthropology blunted the edges of any Carlylean or Huxleyite appeals to rationality that may have emanated from the Royal College of Science. Alternative spaces encountered during his decade in Dublin were to focus his mind on this vision. First, his visits to the Owenite hub that was the Trinity Anatomy Department (see chapter 2) wore down any "Darwinian" appeals to reductionism preached by Huxley's colleagues at the Royal College, and thus tempered biological rationalism with a more holistic study of humankind.

Second, his extensive surveys of Irish marine life while a working zoologist in Dublin were instrumental in underpinning his belief in recapitulation, which ultimately drove his desire to record primitive races before their inexorable extinction. Other progressive evolutionists and recapitulationists marshaled their anthropometric data for political ends, but Haddon's main aim was to record for posterity. Unlike the human anatomists Macalister and Cunningham, Haddon's established specialty was sea anemones. He had published a major revision of British species in which the classification was based on a close consideration of anatomical and embryological features rather than external morphology.[67] Like many of his contemporaries, he experienced no difficulty transferring his ideas on biolological evolution to the study of culture, and his application of zoological order to anthropology is often cited.[68] Haddon's contention—that human cultural life could be placed on a line of development parallel to that of the zoological order, and in both, investigation into the development of the earliest stages was the key to discovering that order—is perhaps the aspect of his thinking that had the most bearing on motivations for his Dublin work.[69]

Haddon was greatly influenced by Francis Balfour (1851–82) and referred to his methods extensively in his 1887 publication on embryology. Through the works of Balfour, Haddon became familiar with the recapitulation theories of Karl Ernst von Baer and their development by Ernst Haeckel. Haeckel's dictum that ontogeny recapitulates phylogeny, an idea accepted by Balfour, was incorporated into Haddon's textbook and transferred directly to anthropology: "The chief result of all embryological inquiry has been to demonstrate that the history of the individual recapitulates in its main features the evolution of the race and thereby to give positive evidence in favour of the theory of evolution in the general acceptance of the term."[70]

Haddon believed that primitive people remained childlike in both physical build and mental ability and this was reflected in their cultural practices; civilized people were truly adult in mind, body, and culture. But civilized children passed through all the stages of evolution on their path to adulthood, and their juvenile games were but a phase on the journey of progress. Within contemporary European cultures traces of earlier primitive forms had survived in folktales, toys, and children's games. All these survivals fascinated Haddon, but taking Haeckel's dictum to its logical conclusion, Haddon could foresee major population extinctions. Crucially, when advanced and primitive races came into contact with each other, the recapitulationist argument held that the primitive races were doomed in the evolutionary struggle. The simpler "foetal" races would be no match for "adult" Europeans. Based on hard data from anthropometry, physical and mental decline was inevitable. Just as in 1862, Joseph Camp, the US statistician and superintendent of the 1860 census, stated that "the gradual extinction of the negro was an unerring certainty," Haddon believed that primitive societies were at greater risk of being lost to science than the marine organisms he studied.[71] In a report to the British Association for the Advancement of Science in 1897, he compared the vanishing of cultures in Britain with the demise of the Pacific Island cultures. The same year in a *Nature* article Haddon subscribed to the view that "savage" customs were fast disappearing as European civilization expanded.[72] His primary an-

FIGURE 14. Alfred Cort Haddon (1855–1940), director of the Anthropometric Laboratory and professor of zoology at the Royal College of Science for Ireland. Reproduced by kind permission of UCD Archives UCDA RCSI/212.

thropometrical motivation in measuring the Aran Islanders, guided by the hierarchical implications of recapitulation theory, was therefore one of *racial retrieval*, rather than *racial differentiation*. This state of affairs left him in an ambivalent position regarding the role of the Anthropometric Laboratory in providing anthropological direction for Home Rule arguments at the time.

As a member of the British Association for the Advancement of Science Committee for the Ethnological Survey of the United Kingdom between 1893 and 1910, he oversaw research that concluded that the population of the British Isles was composed of a mixture of different stocks, and that there were no gross distinctions to be made among the populations of England, Ireland, Scotland, and Wales. These findings were included in his 1910 publication on the races of man.[73] This concurred with the views of other prominent members of the anthropological community such as Lubbock and Huxley, who were of the opinion that because the Irish were not "racially" different in a biological sense, they could therefore be assimilated as full citizens of a parliamentary regime that would negate any need for a disengagement with the Union.

But interestingly, Haddon and Browne made indirect allusion to the possibilities of Home Rule, as they incorporated in their survey of the Aran Islands extensive tracts from the notebooks of Samuel Ferguson, including observations on the South Island inhabitants, which point out, "Here where they have been left to themselves, notwithstanding the natural sterility of their Islands, they are certainly a very superior population—physically, morally, and even economically—to those of many of the mixed and planted districts."[74] What the advocates of National Surveys back in London had not anticipated were the local exigencies of encounters in the field, not least the contest that ensued on Aran between the innate hierarchical presumptions of both Anglo-Saxon anthropometric science and Irish Romantic literary tradition.[75] Haddon then, whether acknowledging an alternative worldview or simply avoiding political confrontation, did not produce an overtly politicized text in the wake of his fieldwork on Aran and his work cannot be easily placed within the debates around Home Rule.

Third, Haddon's interest in geography took his vision of anthropology away from physical anthropology toward regional studies. His teaching stints at the Edinburgh summer school of Patrick Geddes, which took place in 1891, the same summer as the foundation of the Dublin lab, introduced him to the theories of the French social scientist Frédéric Le Play, as well as to Geddes's own fledgling plans to develop closer connections between anthropology and geography and to downplay its links with anatomy.[76] Haddon's first anthropometric report—that on the results of the Aran Islands Survey, which was completed in Dublin—owes much to Geddes's vision of anthropology. On the basis of a classification directly attributed to Geddes, Haddon put physical anthropology on the "lowermost plane," which he termed "Anthropogeography," then came "Ethnology" on a plane higher, and this he saw in terms of the "anatomy of cultural man." By the time he had published the results of the Torres Island expedition in 1903, the highest plane of all was attributed to psychology, which, "while it had its roots in man's nature, flowers so to speak, in a realm of its own."[77]

Fourth, in the early 1890s his plans for the major Ethnological Survey of the British Isles, were still to be implemented. He needed to keep E. W. Brabrook and John Beddoe on board, and continued Royal Irish Academy funding was essential. Any early misgivings about the system could jeopardize his own research. It is in the context of these four issues that Haddon's engagement with the Dublin anthropometric venture can be placed. So, it can be seen that under cover of the extant polygenetic system of anthropometry, with all its inherent racial prejudices, Haddon could gather data to furnish his personal visions of anthropology.

In sum, Haddon's fieldwork contacts of a literary kind, his belief in the mixture of racial stocks across Britain and Ireland, and his geographical discussions ensured a personal interest in the fundamentals of wider human cultural development. This in turn resulted in a political ambivalence when it came to issues of Home Rule, but some leanings in support of home governance were evident. His enthusiasm for the "physicalist" anthropometric mission is best understood, therefore, in terms of its place—as a highly successful experiment—in securing support for future anthropological projects of all kinds.

Cunningham's Habitus

Born and brought up in Crieff, Scotland, where his father was the parish minister, Daniel John Cunningham shared with his predecessor Alexander Macalister a piety characteristic of their common Scottish Presbyterian heritage. However, if Macalister's reservations about craniology signaled a distaste for the polygenetic implications of an essentially "soul-less" enterprise, Cunningham would use the anthropometric data to advance similarly reductive interpretations from a monogenist point of view. Although Cunningham, like Macalister, was an active proponent of monogenism and continued the latter's research into evolutionary linkages between man and apes, his interests revolved around the concept of degeneration.[78] Toward the turn of the century, Cunningham became involved in survey work for the Privy Council Committee on physical deterioration and became a leading advocate of biometrical analysis of anthropological data for race fitness, always being careful to point out the potential bearing of "anthropometric work" on the solution of problems that were engaging the attention of statesmen.[79]

His disciplinary concerns echoed Galton's view that the use of anthropology for human improvement, whether physical or mental, afforded excellent but neglected fields for investigation, but the seeds of such speculation—on the improvement of society through reasoned intervention—were certainly well tended during his academic tenure in a politically divided Dublin. When asked at a meeting of the Inter Departmental Committee about general deterioration in the population, he replied, "If there is any part of the United Kingdom in which physical deterioration is to be expected more than another it is in Ireland." He also drew attention to the "well established fact" that insanity had increased to an alarming extent, quoting that "in 1851 the proportion of insanity and idiocy to the population was 1 in 657 and in 1901 the ratio was 1 in 178."[80] For Cunningham the Dublin Anthropometric Laboratory represented the first tentative steps toward a National Anthropometric Survey, but it had its limitations in being associated with an Ethnological Survey of populations in isolated and outlying areas. His aim was "to collect exact data of the physical and mental characters, the health and environment or conditions of life" of the entire population. In doing this he could furnish his anti–Home Rule arguments with scientific evidence for degeneration in Ireland. In his view, degenerative tendencies could be ameliorated by means of scientific rigor and good governance from London. It was important for Cunningham, therefore, by means of his scientific work and public pronouncements, to draw attention to factors that would seem to indicate that the Irish were incapable of maintaining a home government.

From 1883 to 1903, against the backdrop of the Home Rule debate, Cunningham lived among a population he viewed as being "divided into bitterly hostile sections," the minority of which "contained within itself the very great proportion of the intellect, energy, industry and enterprise of the country," whereas the chief distinguishing feature of the majority was "the very lack of those qualities which serve to distinguish the minority," and he continued, "unless he has learned self restraint by the beneficial influences of education, [the Irishman] is excitable, emotional, superstitious, and, I am sorry to say, in some cases, a dangerous member of society."[81] Cunningham's scientific maxim that "each race had a physical mean or average which it tended to retain and to which individuals reverted whenever adverse influences were removed," can be seen to draw on his earlier politically motivated assertions that "the Irishman [despite his ignorance], as long as he is left to himself, unhindered by "priests" or "Nationalist agitators," is full of "generous impulses,"

FIGURE 15. Daniel J. Cunningham (1850–1909), director of the Anthropometric Laboratory and professor of anatomy at Trinity College Dublin. Wikimedia Commons: Public domain image.

a trait that presumably, if cultivated by good governance, would lead to the betterment of that society.[82]

So, the looming specter of Home Rule can be seen here to have focused Cunningham's anthropological theorizing in the early years of the lab and highlighted the inflection of Dublin's political spaces on the developing science. The closely interwoven connections between anthropological science and local politics were reiterated at an 1893 meeting at Trinity College, where Cunningham proclaimed: "There is, sir, a great natural law which is recognised by scientific men as underlying the workings of all communities of living things. It is called the law of survival of the fittest (Applause). Although in the minority, the Unionists are not afraid to subject themselves to the operations of this law. Let the British power be withdrawn and the test be applied. They do not dare it: nay in the North I rather suspect they court it (cheers)."[83] Another example of the intermingling of specifically Dublin spaces with Cunningham's degeneration message was to occur on a very different plane, this time through his professional contacts with Italianate Catholic spaces in the city. When the enduring Italian tradition of religious art, based on the use of an idealized human body for the transmission of emotion, merged with Catholic revival art in nineteenth-century Ireland,[84] it had the effect of highlighting the "theatrical" in church art styles. This in turn engendered an emotional approach to devotion that filtered down to products intended for a religious mass market. Chief among these manufacturers/suppliers of art for a religious market, was Cunningham's model maker, Casciani and Son, of 32 Wellington Quay, Dublin, whose advertisements announced "the largest stock in Ireland of religious statuary for churches."[85] Casciani also made and supplied anatomical models to clients as far afield as Australia and the Americas.

The plaster anatomical models from Wellington Quay were modeled from cranial dissections conducted by Cunningham in Trinity College during the 1890s. The plaster casts were unusual for their reproduction of the exact facial features of the cadavers. According to Lucinda Spencer, most models from this period aimed to give a dignified face to the human condition, whereas the Casciani versions display "an intriguing combination of technical accuracy and ghoulish realism."[86] By engaging the lab with such spaces, Cunningham ensured that the transformation of the cranial dissections into plaster anatomical models in Dublin carried an emotive and emphatic underpinning of the degeneration message to anatomical teaching collections worldwide. The menace society faced from uncontrolled activities of the feebleminded was clearly discernible in the evocative facial expressions of the Casciani models. Just as Cunningham's Dublin political and religious spaces shaped and informed the wider message of degeneration, his disciplinary spaces allowed him access to physical data that could help corroborate and bolster his scientific theories relating to the same. Two decades were spent gathering anatomical and anthropometrical data to furnish this thesis in Dublin: "It is all very well to talk of physical degeneration, to insist upon the importance of maintaining the high standard of physical excellence which we believe to be the heritage of the British race . . . , such discussions can lead to no real practical or useful issue until they are based on a solid foundation of acquired scientific information."[87] Cunningham's preoccupation with race fitness manifested itself as a lifelong concern with particular variations that displayed signs and symptoms of "atavism" and "epigonism."[88] As he explained to the British Medical Association in 1898, retrospective or atavistic variations "act as signposts planted at very irregular intervals and often appearing at very unexpected places, which indicate the path which the individual has traversed in his ontogenetic and phylogenetic development." On the other hand, prospective or epigonistic variations "afford evidence often most uncertain . . . which give us some inkling of the direction towards which phylogenetic evolution is tending."[89] This research interest was to undergo a major boost with the opening of the new Anthropometric Laboratory.

Cunningham's anatomical research had elucidated that "some parts of the body are much more liable to atavistic variation than others . . . stability of structure is a character of those organs which have a long ancestral history . . . thus, phylogenetically old parts are not, as a rule, liable to variation, while parts which are of comparatively recent phylogenetic origin are very prone to variation." He was therefore confident that "the acquisition of the erect attitude in man, at no distant period in his ancestral history, had profoundly disturbed his whole anatomy and is largely responsible for many of the variations which are so abundantly and frequently met with in his structure," and hence, "the lumbo-sacral region of the spine in man is in a state of very unstable equilibrium and here it is possible to study examples of retrospective and prospective variation." Combining this with his conclusions about "critical periods" in the development of the primate cerebrum, he was able to focus almost exclusively on the cranial and lumbo-sacral regions of the human form in the quest for signs of degeneration.[90] The data sheets in the new Dublin lab provided otpimal feedback for the external measurements of these areas. The polygenetic methodology for differentiating numerically between races as expressed on the Anthropometric Laboratory's data-gathering forms, as we have seen, showed both a marked preference for head measurements to furnish craniological researches and for postcranial measurements to determine stature.

Cunningham continued his anatomical research in tandem with the anthropometric work of the new laboratory, and ever apperceptive of medical anomalies in his routine investigations, a series of publications emerged during the Dublin years that resonated with freakish undertones.[91] Despite calls for a rational, scientific, and empiricist approach to anthropometry, these emotive endeavors served to imbue the Anthropometric Laboratory with atavistic spaces where "giants," "monsters," and "idiots" resided, even before Haddon and Brown began their fieldwork in remoter parts of the country.

Regardless of the insistence on humankind's single origin, Cunningham's belief in degradation from the original prototype precluded an egalitarian classification of racial characteristics. This ensured that concepts of racial inferiority were inherent in his vocabulary despite nonhierarchical scientific aspirations. This was amply illustrated by his attempts to pursue evolutionary investigations into the lumbo-sacral region. Here, it was necessary to cast a net beyond the Dublin morgues to far-flung shores, where anatomical features had been shaped by very different environments: "For more than three years I have made every endeavour to obtain a vertebral column of a native Australian, Negro, Andaman, or Bushman." The Trinity Anatomy Department's subsequent antipodean interludes were crucial in his view, as "in the absence of fresh spines of the lower races, where the lumbar region composed of combined vertebral bodies and intervertebral discs could be examined, proof positive upon the degree and character of the lumbar curve was wanting."[92] The Australian links were to continue during Cunningham's working life, and his connection with William Ramsay Smith in Adelaide proved most prodigious. Ramsay Smith procured and sent organs from Australian aborigines to Cunningham over a fifteen-year period, in his capacity as a physician, coroner, inspector of anatomy, and chairman of the Central Board of Health in the city.[93] However, it was with the help of a friend, T. P. Anderson Stuart of Sydney University, that Cunningham succeeded in securing the spine that he sought for research purposes. The backbone of an Australian girl was sent in a zinc box, "which was packed with great skill and care . . . the curvatures were therefore in no way interfered with and it arrived in perfect condition."[94]

Cunningham used this and Irish material in his collection to develop the "lumbo-vertebral index," from which he went on to demonstrate that the curvature extremes represented by the European and Australian spines were not proof that "the Europeans had assumed the erect attitude at a period antecedent to the low races" but simply a reflection of the relative proportion of cartilage in the construction of the lumbar region. The difference he explained was due to habits, "the European who leads a life which rarely necessitates his forsaking the erect attitude . . . has sacrificed in the lumbar part of the vertebral column *flexibility* for *stability*."[95]

Like Haddon, Cunningham negotiated his way through the polygenetic methodologies of the anthropometric laboratory to furnish his own research interests. His belief in degradation from an original prototype, underpinned by data from the lab, enabled him to advance the monogenetic hypothesis that "the differences in man were owing to the existing diversities of climate and other conditions."[96] But, spurred on by the prevailing political climate in Dublin, he mobilized these findings to warn against the breakup of the union between Great Britain and Ireland.

If acknowledged at all in Dublin lore, the anthropometric episode appears, at best, as an ill-judged local folly—an imperious flurry of social Darwinian scientific racism pitched uncomfortably close to home. However, this chapter has demonstrated that the Dublin laboratory was an intrinsic part of a much wider global crusade to satisfy late nineteenth-century classificatory impulses, that it was situated in the context of broader ethnological investigations emanating from the United States, France, and Britain, and bound up with ideologies of racial origin that preceded the Darwinian intervention.

The chapter outlined first of all how the initial design and thrust of anthropometrics was firmly within the hands of the polygenists, whose methods and concepts became ingrained in the fabric of the discipline. Monogenists could just as readily rank races for political or social reasons, but in scientific terms, their unity paradigm was dismissed until Darwin's challenge. However, even in the late nineteenth century, as monogenists gained the upper hand in the "origin of man" debate, many simply transposed elements of polygenetic thought onto a developmental model of evolution that was more Lamarckian than Darwinian. In this later (recapitulationist) account, living races still formed hierarchies because some were perceived as preserving earlier stages in the development of the highest.

The chapter also explored some of the evidence suggesting that the specific methodologies implemented and instruments used in the Dublin lab were originally designed to highlight these differences of kind, which were absolute, static, and hierarchical. This, it was stated, could be deduced from two fundamental considerations. First, the research aim of the lab—which sought to evince mean or standard numerical dimensions for a priori human types—reflected a polygenetic mindset. And second, the empirical procedures employed— craniometric assessment, facial angle determination, and posture determination—also served to underpin this worldview. But, importantly for studies of late Victorian physical anthropology, monogenetic developmental thought, which retained polygenetic ideas of hierarchy, was equally served by anthropometric results. This brand of anthropological empiricism had a nuanced engagement with political spaces in the late Victorian city. Despite the apparent failure of the British Association's Ethnographic Survey to achieve its stated goals across the United Kingdom, the Dublin lab retained a particularly important place in the Victorian mindset.[97] Spatial probing has revealed how the Dublin siting of the laboratory enabled a local mobilization of the global ideologies of anthropometry for personal and political ends during the Home Rule debate. Further spatial interrogation into the lives of Haddon and Cunningham demonstrated some of the ways in which they negotiated the new biometrical experiment. Although initially commissioned by the lab to pursue racial classification, under its auspices, they instead gathered data to furnish their own visions of anthropology.[98] For Cunningham, this meant the gathering of statistical data "to assess the alleged deterioration of the national physique" and for Haddon it made a valuable contribution toward his quest to understand the "complete history of man."[99]

So, although the aim of the Hibernian antislavery society founded in 1837 by a group of Irish Quakers, as stated by one of its founding members, Dr. James Haughton, was "to put an end to the unholy alliance between Irish men and slaveholders in America," the ensuing American Civil War, which ironically ultimately freed the slave, helped to justify racial hierarchies of nineteenth-century society.[100] The institutionalized attitudes of racial inferiority evident toward the end of the century owed much to anthropometrical methods developed by polygenetic science during the war. James Haughton's founding aim for Irish men to dissociate from slaveholders was realized, but the polygeny that underpinned this same racial subjugation was to resurface closer to home when the Dublin Anthropometric Laboratory was launched on June 25, 1891, in Trinity College Dublin.

Chapter 4

Word

"As a particularly imaginative and expressive use of language, literature unlike philosophy or science, is more suitable for the evocation of national phenomena and the experience of national consciousness."[1] When the Irish philosopher Thomas Duddy made this observation a decade ago in the *New Ireland Review*, he was reflecting the widely held—and still prevalent—perception that the arts generally lend themselves more effectively to the aims of Nationalistic consciousness-raising than do philosophy and science. Literature in particular, he says, can be made to register local landscape, local character, local event, and local story. Philosophy and science on the other hand, precisely because they have recourse to a more abstract language, do not so easily lend themselves to the local, the regional, or the national.

This chapter proposes that far from being removed from the particular and the local, *science* was an intrinsic part of the revival of national consciousness in late nineteenth-century Ireland. The historiography of Irish nationalism, with its major emphasis on the literary revival, has thus far largely overlooked the role of contemporary scientific ideas in shaping the nascent nation.[2] And as a consequence, the counterinfluence of nation building on emerging scientific ideas has similarly been neglected. More generally, this is a theme explored in *Voices of Modernity*, in which Richard Bauman and Charles Briggs point to the central role of language in the modern project and follow the development of linguistic ideologies as ways of promoting and maintaining forms of social hierarchy and creating modern subjects.[3] They have opened up discussion on the ways in which language and tradition have become objects of inquiry, and how these constructions were rooted in conceptions of science, nature, society, and politics. Appraising the issue from this angle, I would like to demonstrate in this chapter some of the ways in which the sciences of comparative philology and evolutionary biology coalesced with the Nationalist literary endeavor in late nineteenth-century Ireland.

Central to the argument is the contention that any historiographical attempt to critique this Irish-language revival—which conceptualizes "literature" solely as an "expressive use of language" for the "evocation of national phenomena"—loses under its radar those contributors who traced graphemic maps to the new nation and deciphered its phonetic codes. In addition, efforts to recall the experience of national consciousness that converge only on the emotive assertions of language miss what those same utterances reveal—perhaps more tellingly to a post-Darwinian Nationalist—about the inner workings of the orator's skull. By attending to these less-well-aired facets of inquiry, scientific ideas are demonstrated to have been inextricably linked to the late nineteenth-century Irish-language revival in Dublin. And on the other hand, the nationalist endeavor is shown to have indelibly marked aspects of contemporary science.

These processes are explored by examining some of the epistemic spaces of the Nationalist activists George Sigerson (1836–1925), a scientist and president of the National Literary Society (1893–1925), and Eoin MacNeill (1867–1945), a historian and cofounder of the Gaelic League. Based on these two scholars, I show that the reception and mobilization of theories—and their application to language—varied considerably within Nationalist spaces and were critically dependent on the nature of the savant's engagement with the history and politics of nationhood. The Gaelic Revival is set here in the context of contemporary European Romanticism—and its converging ideas of language, ideology, nation, and identity.

R. H. Robins explains that as Europeans became more aware of language diversity in the nineteenth century, projects of new universal languages or attempts at "debabelisation" were just another "swing to the universal that coincided with the elevation of the objectivity and rationalisation of scientific thought and technological progress," an observation underlined by David Harvey when he speaks of the nineteenth-century belief in linear progress, absolute truths, rational planning of ideal social order, and standardization of knowledge and production.[4] But just as scientific technologies and economic progress forged ahead, critical voices such as that of Mathew Arnold condemned the sacrifice of human culture to mechanism and technology. Eric Hobsbawn made reference to this late eighteenth-century countermovement, when he declared that "Europe had been swept by the Romantic passion for the pure, simple and uncorrupted peasantry, and for the folkloric rediscovery of the people and their vernacular languages."[5] In sum, attempts to link language and ethnicity emerged largely in opposition to contemporary European themes of universality. The first stirrings of thought in this direction asserted the inseparability of language and thought and can be traced to the works of Johann Gottfried Herder (1744–1803), particularly his 1772 essay *Abhandlung über den Ursprung der Sprache*, which proposed the idea that every culture and every age has its own unique character, its own distinctive way of thinking and experiencing, and, crucially, its own distinctive language. Wilhelm von Humboldt took up the argument with the claim that the differences between languages lie not merely in sounds but in a people's attitudes and understanding of the world; it was, he believed, "the outer appearance of the spirit of a people; the language is their spirit and the spirit is their language."[6] Numerous authors also draw our attention to the fact that the "essentialist" origins of anthropology lie not only in the nineteenth-century quest to find biological correlates for race but also in the new forging of a close ideological connection between language and identity.[7] In this version of ethnicity, identity is rooted not in genetics, but in heritable cultural forms, especially language, which can embody an ethnic group's distinctive cultural identity.

Bounding the nation-state became one of the preoccupations of European thought during the nineteenth century. In Ireland the founding of the National Literary Society in 1892 and the Gaelic League in 1893 coincided with the defeat of the second Home Rule Bill, and P. J. Mathews draws our attention to the new generation of Irish Nationalist intellectuals who devised strategies for a form of de facto Home Rule. With the realization that little could be achieved at Westminster, the plan was to mobilize and apply the latent national intelligence of the country to the practical needs of Ireland, a strategy encapsulated in the term "self-help."[8] There was a consensus among the revivalists that a more noble representation of Irishness was needed to displace colonial stereotypes. Central to any debate within nationalism was the matter of what should constitute authentic "Irishness." The correlation between language and cultural identity in Ireland and the subsequent drive toward independence were exemplars of broader attempts across Europe to establish "imagined" communities.[9] Although the Irish revivalist programs of Nationalists such as George Sigerson and Eoin MacNeill were situated well within this discourse, their conception of the "imagined" community diverged. With this in mind, and to illuminate how scientific interplay with projects of language and identity was differently mobilized, constrained, and enabled by these different "imagined" communities, it will be wise throughout the chapter to take heed of Mathew's suggestion that the question is not which conception of Irish identity represents "authentic Irishness," but rather "how various notions of Irish identity came to gain ascendancy, to be sanctioned, legitimised, and challenged in a nation that had not yet found full political expression."[10]

During this period, two visions of Irish society and contrasting places for the Irish language in it were crystallized. The Literary Society was founded with the aim of creating a new Irish literature, and it was the vehicle for one vision. This view of Irish society saw Gaelic Irish literature and language as an inspirational source for an Irish literature in English, or Hiberno-English. The most innovative exponents, Yeats, Synge, Lady Gregory, and George Sigerson developed it as a dramatic linguistic form based on the literal translation of Irish into English. This aspect of the revival engaged the Anglo-Irish intelligentsia and the Gaelic Irish learned tradition, while idealizing the Irish peasantry and the rural west of Ireland. In this way, it can be seen that much of the scholarship of the Literary Society underlined the role of the Irish language as a means of accessing the past.[11]

In contrast, an alternative view of Irish society as promulgated by the founders of the Gaelic League in 1892 identified the Irish language and its revival as a spoken tongue as being central to the identity of a new modern Ireland. The vision was captured by Douglas Hyde in his speech on the necessity for the de-Anglicization of Ireland, and one of the aims of the movement was to create a popular literature in the Irish language.[12] The movement also proved successful in influencing public policy. Its campaign to establish the Irish language as a compulsory subject for matriculation to the new University of Ireland was ultimately successful. A Romantic understanding of language "tied it to the spiritual essence of its peoples," and as a result, languages, like the cultural identities that gave rise to them, were thought to be "necessarily separate and non-overlapping."[13] This critique by Mary Bucholtz and Kira Hall offers some insight into the contrasting place of language among Irish Nationalists. If we propose that the National Literary Society savants were consumed by the "spiritual essence" of cultural identity in of itself, and the Gaelic League scholars were guided more by the political potential of the "separateness" it engendered, this bifurcated Irish Nationalist model sits well with contemporary but also dichotomous European intellectual movements of the nineteenth century.

In the first instance, the linking of language to the "spiritual essence" of a people gave rise to what Sinead Garrigan Mattar has described as a "Romantic primitivist tradition," which, combined with antiquarianism and the popular "Celticism" engendered by James Macpherson's notorious eighteenth-century translations of Ossian, was manifested in works as such Ernst Renan's essay "La poèsie des races celtiques" and Matthew Arnold's *On the Study of Celt-*

ic Literature. The noble savage was the most cited example of this Romantic primitivism, and in these Romantic readings early societies such as the Celts, unearthed by antiquarian pursuit, were projected as a passionate, isolated, melancholy, creative, and self-sufficient race.

As the rise of anthropological science in the nineteenth century began to dispel the notion that "primitive" peoples were in any sense exemplars of an Edenic purity from which Western society had fallen, the degenerationist model behind the noble savage was transformed into an evolutionary one, in which "primitive" was conceived as alien to the civilized mind. This reading characterized "primitive peoples," the Irish included, as a stark case of arrested development. The first task of the Irish Literary Revival was thus to dismantle such negative images, particularly those made popular by Charles Kingsley along with *Punch* and other English comic weeklies. They sought instead to invert the stereotype and valorize the peasant as a "spiritual figure, the living embodiment of the Celtic imagination, a natural aristocrat."[14] The folklore and imagination of the idealized peasant were pitted against the modern industrial and commercial British spirit. As Nicholas Allan points out, this was one of the ways in which Irish Nationalists adopted evolution as evidence of their nation's unsuitability for union with Britain. He highlights the role of evolution as an enabling discourse for the Irish Revival.[15] With this in mind, the Literary Society and Gaelic League are examined here as epistemic sites that shaped the discourse on language origins in different ways.

First, George Sigerson's and the Literary Society's projection of early Irish society adhered to a traditional Romantic mode, denying that the primitive Irish were anything less than noble "Milesians" with mysterious origins. When the translation of seventh-century Irish texts such as *Auracepit na Eces* revealed attempts on the part of early Irish grammarians to defend spoken Gaelic over learned Latin, and European scholars identified Celtic languages as providing the closest links to the Indo-European *Ursprache* or parent language,[16] the Literary Society Nationalists" belief in a "civilised" past was affirmed. The possibility that Gaelic was one of the candidates for a living descendant of the Adamic language, and the subsequent argument that "this first born and consequently supernatural language retained traces of its original isomorphism with the created world" was grist to Sigerson's mill.[17] Second, as the anthropological sciences were encroaching on the "spiritual essence" of Romanticism and ruffling the sensibilities of Literary Society Nationalists, so too, was biology making incursions on its linguistic mechanisms.[18] The emergence of comparative philology heralded a new scientific era in the approach to language study, and its concerns ran the gamut from "organicist" speculation to neogrammarian "atomism."[19] As a consequence, the study of "Celtology," which had hitherto been dismissed by science as an eighteenth-century Romantic fabrication, was given new authority—particularly following Franz Bopp's inclusion of Celtic in his comparative grammar of Indo-European languages.[20] This and other contemporary comparative studies of old Irish attracted the attention of the Gaelic League Nationalists, as did the modernizing potential of the science that underpinned them. So, from the perspective of the Nationalist language revival, the pursuit of the "primitive" was becoming polarized into spaces of "science" and "non-science" in Dublin.

This chapter argues that Sigerson's and the Literary Society's approach to the Irish language was intimately connected with an organicist discourse grounded in late Victorian evolutionary biology, and in particular, the language–species debate and its concomitant human–animal linkages.[21] The crucial influence of the German linguistic scholar Max Müller (1823–1900) is highlighted, as is his epistemological orientation, which derived from Kant's *Critique of Pure Reason*. This approach was characterized by a belief in biblical monogenism, common descent, and the insistence that language and thought were inseparable. On the other hand, it suggests that the Dublin spaces of Eoin MacNeill placed his analysis of the Irish language well within emergent methodological programs in European science. The strand of comparative philology that interested MacNeill was less concerned with organic interspecies links or, indeed, with links to an original mother tongue. His systematic approach derived from the German scholars Franz Bopp (1791–1867) and Jacob Grimm (1785–1863), and was driven by a pressing need to establish a typology of fundamental linguistic stocks. Franz Bopp declared it his purpose to give a "more scientific" treatment of language, one that would trace the natural-historical laws "according to which occurred its development or destruction or rebirth from previous ruination."[22] This approach culminated in the neogrammarian movement, which rejected historicist beliefs about linguistic growth and decay. The important point here for a scientific critique of Irish Nationalist language studies is the differentiation between one nineteenth-century analytical style that remained resolutely historical in its approach, and another that adopted a synchronic approach and viewed the language primarily as a system in its *current* state, rather than in its evolution.[23] As we shall see, Sigerson was concerned with translating into English, textual material from the Gaelic past, whereas MacNeill sought to describe the language as a total system and to characterize phonological, morphological, and grammatical regularities irrespective of their antique or modern status.

So, from the moment Eoin MacNeill's scholarly interests turned to focus on the rudiments of Gaelic, a fault line appeared through the investigative spaces of the Irish revivalist terrain. As the Gaelic League and National Literary Society further expounded their visions of the new nation, the chasm deepened. Interpretations of the "primitive," of science, and ultimately of identity and nationhood appealed to an authority informed largely by either an idealist or mechanistic viewpoint. Being alert to these observations, something of the ways in which science was molded to enhance the linguistic ideologies of Sigerson and MacNeill will be outlined, and in so doing, will also reveal how a selective use of contemporary scientific motifs flourished in different ways in different Nationalist spaces.

Language and the National Vision of George Sigerson

The chief secretary to the lord lieutenant of Ireland, Colonel King Harmon, was asked in a House of Commons debate in 1884 if he was aware that a certain Dr. Sigerson of Dublin—recently appointed to a position of trust and influence by her Majesty's Government—"was the writer of many articles of a seditious nature in the Irishman newspaper." A Dr. Lyons quickly interjected that he was the same gentleman, "whose researches into the physical condition of the atmosphere had been described as a work of *great authority* in Parke's history of Hygiene, and for the Hon. and gallant member not to know Dr. Sigerson was not an argument that he himself was unknown."[24]

George Sigerson (1836–1925) was a prominent Dublin public figure, celebrated as a Nationalist poet, patriot, scientist, and scholar, but remembered particularly for his thirty years as president of the revivalist National Literary Society. Born at Holyhill, Strabane, County Tyrone, on January 11, 1836, he attended school at Letterkenny Academy and St. Joseph's College, Montrouge, in Paris. There he excelled in Latin, Greek, German, French, and literary analysis, before beginning his medical degree at the Queen's University of Ireland in 1859. He also studied at the Catholic University Medical School on Cecilia Street, Dublin, with which he was to have a long academic connection. Later, as a professor of biology at the National University of Ireland, Sigerson lectured medical students in both botany and zoology. In 1883 the remnants of the Art and Science Faculties of the Catholic University were entrusted to the Jesuits, who set up University College, Dublin, to train students for examinations of the Royal University of Ireland (established in 1879 as a purely examining institution). Sigerson was appointed professor of biology in this new college from the outset, and consequently was to teach biology in two historically linked but now independent institutions.[25] That George Sigerson was noted in Westminster in the 1880s for environmental issues relating to his "atmospheric investigations" and for his social concerns, particularly "the special attention given to hospitals and matters connected with public health,"

FIGURE 16. George Sigerson (1836–1925), physician and lecturer at the Cecilia Street Medical School and president of the National Literary Society. Painting attributed to Sir John Lavery (1856–1941). This image is reproduced courtesy of the National Library of Ireland, PD SIGE-GE(1) IV.

would suggest that the recent invigoration of his reputation as Dublin's first neurologist, was not generally acknowledged in nineteenth-century scientific circles.[26]

By placing the investigations of George Sigerson within a Romantic scientific tradition rooted in the eighteenth century, this study runs counter to recent moves to center him within the modernist thrust of the revivalist movement, which draws on a specifically Irish conversation between science, modernity, and national identity.[27] Instead, Sigerson's continental brand of biological thought, is shown to have underpinned the Literary Society's revival of the Irish language. For in fact, both Sigerson's literary and scientific ambitions were driven by a Romantic conception of life, in which attempts to understand animate nature, through poetry, philosophy, and science, were foregrounded over the new professional and specialist science making an impression in Dublin's medical and academic institutions of the time.

That contemporary medical opinion in Dublin placed him outside the bounds of progressive medical science is further attested by his failure to obtain a hospital appointment in Dublin—a prerequisite for admission to the medical elite. This is all the more noteworthy given the close connections between the Catholic University School of Medicine and its religiously associated St.Vincent's, Jervis Street, and Mater Hospitals, and cannot be seen in terms of religious discrimination.[28] J. B. Lyons ascribes this failing to Sigerson's involvement in "political journalism," and although holding academic attachments in botany and zoology, he notes that Sigerson's first post was "in that unendowed and struggling institution."[29] Notwithstanding later professional accolades, George Sigerson had to rely on his clinical practice to produce an adequate living. The following pages examine Sigerson's habitus—a merger of Romantic idealism, Literary Society associations, and Parisian connections—to demonstrate how these spaces largely directed his experimental work and publications in Dublin.

His medical practice at No. 3 Clare Street was close to the East (or back) Gate of Trinity College. Nearby were the college medical and science buildings, and the Lincoln Place tea shop, where George Frances Fitzgerald held forth with his colleagues on the problems of electromagnetic theory (see chapter 1). The scientific approach that most directly linked with Romanticism—that which foregrounded ideal plans in nature—was generally thought to have perished at the hands of late nineteenth-century figures such as Thomas Huxley and Hermann von Helmholtz. It resisted however in Dublin's fashionable quarters, this time beyond—if only just—the walls of Trinity College. From this angle, it can be seen that Sigerson's historiographical placement as part of Dublin's nineteenth-century professional medical establishment

FIGURE 17. Cartoon of George Sigerson by Grace Clifford (1916). This image is reproduced courtesy of the National Library of Ireland, PD2159TX (16) 59.

is largely due to the proximity of his residence—and clinical location from 1877—to the Merrion Square epicenter of eminent practitioners such as William Wilde, Dominic Corrigan, and William Stokes, and not attributable to local scholarly recognition by the late Victorian Dublin medical elite. The Romantic mode of being and thought, as discussed by Robert J. Richards, David Knight, Andrew Cunningham, Nicholas Jardine, and more recently, John Tresch, was embodied in Sigerson's persona.[30] As Richards tells us, "We catch ideas in the making only when we understand rather intimately the character—the attitude, the intellectual beliefs, the emotional reactions—of the thinkers in question."[31] Medical expertise notwithstanding, Sigerson's "French éloge allied to Irish wit"—and a "great mane of silver white hair, resembling a Sigur's son"—positioned him as clinician of choice to the prominent literary savants of Nationalist Dublin.[32] Contemporary accounts of his Clare Street home make much of the Sunday salons, where lobster, goose, joints, and puddings were enjoyed in the "Balzac-like interior" of French furnishings, paintings, pastels, engravings, and sculpture. Literary figures such as Yeats and Hyde conversed here with Trinity academics and distinguished visitors to Dublin. Dinner party discussions at No. 3 occasionally turned to the occult, and "table tapping" was undertaken by guests in a room at the front of the house.[33]

Contacts of a pagan, supernatural, and theosophic sort were also to be had within the walls of the National Literary Society's meeting rooms at No. 6 St. Stephen's Green.[34] Here Sigerson presided over a revivalist literary community that combined notions of a glorious Gaelic past with a fundamental Romantic tenet—the original unity of man and nature in a Golden Age. Although the society welcomed all comers, membership lists indicate a predominantly middle- and upper-class following, with the majority of registered addresses coming from Dublin's fashionable suburbs of Rathmines, Dalkey, Foxrock, Clontarf, Malahide, and up-market city locations.[35]

The society itself was founded by William Butler Yeats and John O'Leary on May 24, 1892, with the aim of developing and fostering the intellectual resources of Ireland. The main objective was ostensibly to publicize the literature, folklore, and legends of Ireland, but also to create a market for the literary intelligentsia. With a high level of support from interested men of letters who were attracted to the nonpolitical nature of the movement, it became possible to plan various projects and schemes, including cultural agitation through lectures, concerts, conversaziones, and poetry readings; the establishment of a network of Nationalist reading rooms throughout the country; and the mass distribution of a series of new Nationalist books. The society sought to exert as much influence as possible on national opinion and national taste by encouraging Irish readers to look on Ireland rather than England as the focus of cultural value.[36]

This aspect of the revival was not without its Nationalist detractors. In 1899, Eoin MacNeill remarked caustically that "the so called Irish Literary Movement is a hindrance not a help to a genuine revival" and he also derided those who "were lying in wait for the latest piece of Gaelic lore dug out by the philologist, to dress it up in an alien garb, and turn it into third rate English poetry." Other Gaelic League members thought that the Literary Society's idea of heroism was more pagan than Christian and that it taught young people to "regard their country not so much as a thing they should be satisfied to serve" but rather as "a stage on which they may expect to play a part in the drama of heroism."[37]

With the founding of the Literary Society in Dublin, it was Yeats's intention in part to create a standard of criticism for Irish material, and he duly set himself up as a self-styled arbiter of taste. His intellectual influence on the output of the Literary Society was therefore seen as immense. Simon Naylor has detailed the relationships of institutions with particular associated characters and has shown how an almost symbiotic relationship developed, one informing and reinforcing the other, the union often becoming disproportionately influential in the advancement of ideas.[38] In this way, it is possible to envisage Yeats's negative experiences of Darwinism, as the momentum that steered the editorial and critical output of the society, and thus colored disproportionately its resistance to contemporary science—a theme common to revival historiography. However, on closer inspection, it can be seen that the Literary Society did embrace a scientific worldview, but it was cloaked in the Romantic biology of its other powerful persona, George Sigerson, and was of a sort that would not survive the positivist onslaught of the new century. However, the influence of Sigerson's science on the Literary Revival has yet to be explored. To begin with, if we reel back to the first edition of the *Literary Society Journal* of 1900, and the claim made by its prefatory note, that "no nation can be disarticulated from its past without paralysis of some of its highest powers . . . the spiritual, intellectual and chivalric qualities so characteristic of the ancient nation, should be made prominent as of old,"[39] our attention is drawn to the biological metaphor, which hinted at the interrelated nature of Sigerson's spaces—Romantic nationalism and cerebral anesthesia.

The belief that Irish folklore and language in some way maintained the esoteric truths that were once part of a world religion, a revealed truth that might be reclaimed, was central to the fabric of the National Literary Society. That these ideas, which provided a worldview alternative to the one propounded by Huxley and John Tyndall, were given serious scientific consideration by Literary Society members, reflects the persistence and persuasiveness of Sigerson's Romantic biology in late nineteenth-century Dublin.

Sigerson's literary bent, Nationalist connections and religious beliefs all channeled into his scientific outlook, and provided him with views considered unorthodox in an increasingly specialized, professional Dublin medical community. He disputed the materialist position that the belief in the existence of God and the immortality of the soul was unknown in the early generation of mankind and only dawned on humanity gradually: "The analogy of the physical order does not deny but advocates that in the moral order man became immediately cognizant of some great central eternal truths."[40] This theological concept of an original mental complexity was transposed to his vision of Irish origins and profoundly affected the way he understood and mobilized evolutionary theories. In an 1893 address on the revival of Irish literature, he states: "And it may be a comfort to know, in view of prevalent hypotheses, that the stock of the anthropoids never went through evolutions in this country. Whatever may have happened elsewhere, the beings who first leaped upon our shore must have been among the foremost in the developed attributes of manhood." And again: "It is certain that intellectual cultivation existed in Ireland long before the coming of Saint Patrick. We have the laws of the revision of which he assisted, and I assert that speaking biologically, such laws could not emanate from any race whose brains had not been subject to the quickening influences of education for generations."[41] The following year in the *New Ireland Review*, a popular Catholic literary monthly, Sigerson took up the argument against a materialistic evolution with the following observation: "If belief in a Divine being and an undying soul were false, then the progress of the human race would have resulted in reducing us mentally far beneath our animal ancestors . . . no, the hypothesis of the materialists cannot stand. There has been no progress, or progress corroborates faith . . . there is no escape from this dilemma; either the doctrine of development and progress is radically false or the understanding of Moses was enlightened by knowledge beyond the ken of Mortals." He singles out for particular criticism Ernst Haeckel's naturalistic pronouncements: "It did not manifestly occur to him to wonder how a member of the yet but half developed human race could have conceived or expounded the grand ideas which attract his admiration . . . the apostle of a new science finds it to abound in majestic scientific truths."[42]

The views emanating from both the Anglo-Irish and Gaelic Irish strands of the revival were loosely tethered by an adherence to the theological and philological ideas of the nineteenth-century German Sanskrit scholar Friedrich Max Müller. Müller, credited with giving the science of philology a public status in his *Three Lectures on the Science of Language* (1864), denied that a belief in evolution necessarily negated the possibility of revelation. He was profoundly influenced by the Kantian transcendentalist model of spirituality and was opposed to Darwinian ideas of human development, arguing that "language forms an impassable barrier between man and beast."[43]

Müller's mingling of language and theology was largely greeted with enthusiasm in another of Sigerson's intellectual spaces—Catholic revivalist circles.[44] Here Canon Ulick Bourke had popularized much of the philological research of continental scholars such as Franz Bopp and Johann Kaspar Zeuss for Irish readers. Bourke found in Müller's *Science of Language* series an appropriate framework on which to hang nascent theories of Gaelic revivalism, and pin Irish origins on a heroic Celtic Aryan past. Bourke was a biblical monogenist and used many of Müller's arguments to advance this stance. Drawing on archaeology and linguistics, Bourke's 1875 publication *The Aryan Origin of the Celtic Race and Language* allied an emerging nationalism with a literal biblical monogenism in the Gaelic revival debates by arguing that the Gaelic language and people had come from the cradle land of the Japhetic race.

Empirical evidence for the antiquity of the language was gleaned by Bourke from the findings of recent European philological scholarship, and he strove to disseminate as widely as possible revelations that "Irish Gaelic can supply every form of articulate sound with twelve consonant and five vowel sounds by the use of the slender and broad vowels and the process of aspirating and eclipsing," and "there is no form of palatal, guttural, labial or nasal sound, slender or broad that cannot be readily articulated by an Irish speaker." He claimed that such research had shown that Irish—with its sixteen or seventeen primitive letters—had an earlier start westward than either Greek or Latin from the Aryan region. These views encouraged students of Gaelic to examine the Irish language in the context of a biblical framework that saw the original inhabitants of Ireland as having come from the east with an Aryan language; the Aryan region of their origin being "that high table land around Mount Ararat where the four rivers flowing out of the Garden of Eden took their rise."[45] Sigerson interjected here and used contemporary science to continue the narrative of Irish origins propounded by Bourke.

It seems that despite his admonition of certain evolutionary ideologies, he ultimately found his own developmental accommodation. As he explained, "Human skill in particular departments may ascend progressively till it reach its zenith and then gradually decline. Mankind acquires, but loses also; an advance in one direction may mean retreat in another."[46] So, through a careful study of the history of the Irish language, he deduced evolutionary adaptations of the auricular and auditory components of the early Irish cranial and postcranial skeleton. On classic Irish verse, he commented, "Its existence betrayed a highly refined development of the organs of speech and of hearing" and he continued:

> Besides possessing the sounds we recognise and those which other European nations have noticed, the ancient Irish composers noted, identified, and employed other and more subtle shades of sound . . . and this has a physiological as well as literary interest, the human ear had not then acquired the power of distinguishing and taking pleasure in these sound echoes or repetitions which we call rhymes—there is an entire absence of rhyme from the classic composition and from the Gothic dialects in the early ages, and therefore we must infer that the producers were deaf to the nice distinctions of chiming sounds, in other words rhyme deaf. . . . Granting that the ancient Irish possessed the gift of discerning and composing rhymes before other European Nations, as well as highly developed metric machinery . . . and with the ancient language is passing away some of the articulation gains as with our ancient civilisation have disappeared some of the educated powers of eye, ear and hand.[47]

Like Müller, Sigerson did not reject evolution, but denied that a belief in evolution necessarily negated the possibility of revelation. His reading of Darwin's theory was molded to fit a nonmaterialist understanding of evolution that stopped short of the human intellect and embodied an ethical dimension that underpinned a belief in a common human ancestry. Using Müller's research into the common origin of all languages as testimony, Sigerson outlined his understanding of Darwin's theory in an article titled "Genesis and Evolution" as follows:

> The new scientific theory of which Darwinism has proved of immense advantage—an advantage I noted long since when first the hypothesis came under my observation. For a considerable time, down to recent days it was held that there could be no variation in species, none in the several races of man. They were stereotyped. Now as there were many and different races it followed necessarily that if these races were immutable, these races must have had a distinct origin, and its own primal parents. Hence also there must have been many creative centres—a separate Eden for every race of man—laid down by Voltarian school as insurmountable obstacle to the acceptance of the Christian belief in the unity of the human race—whoever did not attorn to this view was regarded as non-scientific and trammeled by unscientific bonds—this impregnable theory has vanished like an ice berg because . . . the evolution theory has destroyed it and proved that all races have sprung from common parents.[48]

The spaces of Adamic monogenism absorbed by Sigerson in Catholic theological circles were bolstered by spaces of "mutual aid" and commonality inherent in the Nationalist vision of his social reforms. This term was coined by the Russian evolutionist Piotr Kropotkin, who rejected the idea of centralized state systems and the social Darwinian stress on competition evident in progressive ideologies. He advocated instead small self-governing, nonprofit-making communities based on "mutual aid" and "co-operation."[49] Sigerson took the opportunity to draw these ideas into his political spaces too. In a piece from his 1868 book *Modern Ireland: Its Vital Questions, Secret Societies and Government*, he derided the distinction between Celt and Anglo-Saxon and in a preface to the later *Last Independent Parliament of Ireland* claimed that "since the first dividing pale there has manifestly been in Ireland a great vital fusing force which attracted all new comers into the nation."[50] This desire to find unity and cooperation rather than difference among races, led him to view Darwin's theory as ethically commendable on the one hand, while strengthening his Christian convictions on the other.

St. George Jackson Mivart had paved the way on this matter for Catholic biologists in 1874, when in response to materialist support of theories put forward in Darwin's *The Descent of Man*, he had claimed that "we must entirely dismiss . . . that the conception that mere anatomy by itself can have any decisive bearing on the question as to man's nature and being as a whole." He embarked on extensive investigations of primate anatomy and concluded that "however near to apes may be the body of man, whatever the kind or number of resemblances between them, it should be always borne in mind that it is to no one kind of Ape that man has any special or exclusive affinities."[51]

Mivart was advancing such empirical anatomical evidence to encourage Catholic biologists to situate their studies within an intellectual framework cognizant of the duality of matter and spirit, and Max Müller's linguistic studies were appealing to a Catholic mindset by looking to Sanskrit texts as the first utterances of a generic being—the first mental products of a spiritually unified humankind. These local Catholic spaces harmonized well with the thesis emphatically underscored by the Dublin literati—that humans in their primitive state were the vessel of revealed religion, and not the first link in an evolutionary chain. The confidence with which this spirit of Romantic organicism was propounded by the literary clientele may well be traced back to Sigerson's French experiences. Here, reputable organs of the French scientific establishment had put their authoritative stamp on scientific practices, which were to prove useful for this Nationalist narrative. As a result, Sigerson's neurological research was to find more fertile expression away from the grip of establishment medicine in Dublin.

Throughout the 1870s, while he was attached to the Catholic University Medical School, Sigerson was a member of the Clinical Society of Paris, and the Parisian Society for Physiology and Psychology, and he traveled regularly to France to collaborate with Jean Martin Charcot (1825–93), DuChenne de Boulogne (1806–75), and Claude Bernard (1813–78). His early medical interests had converged on the human nervous system—the locus of language, abstract representation, and transmission of culture—and these concerns were further awakened by Parisian "hysteria shows" at the Salpêtrière, where novel clinical demonstrations and patient interviews by Jean Martin Charcot had been replacing more traditional hospital rounds.[52] This "Napoleon of the neuroses,"[53] as he had become known, had in the late nineteenth century established and popularized the discipline of neurology using unconventional methods such as hypnosis.

For Sigerson, one of the main appeals of Charcot's approach to studies of the nervous system, lay in the fact that he supplied rigorous scientific and medical explanations to hitherto inexplicable forms of social and religious behavior.[54] This intellectual trend provided Sigerson with the scientific authority for exploring more holistic ideas about the relationship between humankind, nature, and language that were unavailable to him closer to home. Alfred Binet, a student of Charcot, remarked in 1892 that "the researches of the Salpêtrière school served as the point of departure for a *fresh* scientific movement which continues up to the present day."[55] Following a visit to the Salpêtrière in the 1890s, George Roberson, an eminent physician from Edinburgh's Royal Asylum, commented that "the most contradictory statements on matters of observation had been made by distinguished physicians, and phenomena had also been recorded, which, if true, seemed to overturn the established laws of nature and all our fixed ideas of things."[56]

Anne Harrington's study has revealed the complex relationship that had grown up between Parisian neurology and various shades of mesmeric thought throughout the nineteenth century, and in particular, its fin-de-siècle rapprochement.[57] Sigerson, the physician, was thus drawn to Paris at a time of experimentation, when the city's foremost neurologists were engaging in what Marshall describes as "the contradictory exchange between formal neuropsychiatry and parascientific models of the body and self."[58] Despite Jan Goldstein's political interpretation of the work of the Salpêtrière as the "clinical" arm of the anticlerical Third Republic government, which sought to systematically demystify miracles and other such marvels, Charcot's considered engagement with hypnosis can be traced back to the 1870s, when both he and Sigerson jointly investigated phenomena hitherto dismissed as unscientific.[59]

These developments unfolded in 1876, when the very reputable Société de Biologie, a French scientific organization "conceived under the impulsion of the positivist spirit,"[60] finally accepted an application by an elderly French physician Victor Marie Burq, to present the findings of his life's labor on the little-known practice of "metalloscopy." Dr. Burq was an advocate of animal mesmerism and had been experimenting for years on the therapeutic possibilities of metals and magnets on the human mind and body. Sigerson's Parisian reputation as a "man of science" was thus enhanced that year by his being invited to join a commission, along with Charcot, to investigate these neuro-physiological phenomena. Sigerson, who had already experimented in 1869 on the physiological effects of exposure to metals and toxic particles in ambient air, readily took up the appointment.[61]

The investigations made by the group into Burq's metalloscopy suggested that certain aesthesiogenic agents could be used to both inhibit and provoke trance states in the hysterical nervous system. As Harrington explains, this gave rise to the idea that hypnosis could be induced and broken using a wide range of physical agents that worked by irritating the peculiarly sensitive nervous system of the hysterical patient through action on the sensibility.[62] In Sigerson's words, "It has established that certain important results do follow the application of metals, what was first regarded as a chimerical fancy has been proved to be a fact capable of verification by the work of independent observers."[63] Sigerson's reports on the Commission's observations—on Burq's findings and on subsequent similar animal experiments, by Professor Schiff of Germany, effectively to eliminate "the agency of expectant attention"—were published in the *British Medical Journal* in February and April of 1879.[64] In sum, the results of the investigations convinced the French medical establishment of the reality of hypnotic states and endorsed the biological validity of the subject's responses to metallic, electromagnetic, and hypnotic stimuli that had previously been attributed to nothing more than patients' ignorance, imagination, or fraud.[65] "Pathology," Charcot declared to his audience at the Salpêtrière in 1878, "has the right to observe even those phenomena which cannot at once be explained by existing facts or theories . . . even though their novelty . . . may give rise to hostile comment."[66]

So, under the new banner of hypnosis experimentation, investigation into the physiological effects of metals and magnets remained a fundamental part of the research program of the Paris school. Positivist science, therefore, interacted with and gave authority to the practices of metalloscopy and hypnosis, but in so doing, also connected with a wider culture of bodily investigation. The mind and body became a composite receptacle for information, patterns of electrical and vibratory forces, and other fluid physical properties. This culminated in theories such as those of Alexandre Barety in 1881, on the existence of a *force neurique*, and of Alfred Binet and Charles Féré on "physic transfer" in 1892. As a result of the resurgence of this notion of a vital human principle, the clinic investigations, as Harrington and Jonathan Marshall both allude to, became the stalking horse for theosophists, occultists, and mesmerists. Sigerson's French connections thus became a crucial conduit for the Literary Society's scientific underpinning of the Nationalist narrative. By 1893, influential revivalist publications such as the *Celtic Twilight* had referenced the "dark powers that cling about us" and had shared with the Dublin savants the notion of "communicating in states of trance with the angelical and faery beings."[67]

For Sigerson, the Parisian experiences opened up a vista of inquiry that allowed him the space to broaden his research into the workings of the brain in order to counter emerging Darwinian theories of lateralization, which were beginning to question the belief that only humans had the propensity for articulate language. Sigerson's quest reflected a Romantic disposition in which comprehension of the whole had to precede that of the parts. For him, scientific investigation would provide clues back to a Golden Age "in which the lost unity of self and nature might be restored."[68] This had particular ramifications for the type of nationalism propounded by the Literary Society, especially when viewed in the light of the contemporary notion of degeneration, which had become manifest in a series of medical, moral, and cultural crises throughout nineteenth-century Europe. In his widely read and controversial book *Degeneration* (1892), Max Nordau lamented increasing rates of hysteria and mental disorder and their reflection in "degenerate" cultural forms.

Anthropologists viewed madness as closer to a primitive animallike state than to human civilization, and many viewed entire peoples as degenerate in having fallen from the true properties of their racial type through exposure to foreign climates and cultures. The idea conjured up a kind of collective genetic decay variously plaguing a specific nationality or the entire human species.[69] This mood, prevalent in Irish Nationalist circles, was summed up succinctly by Douglas Hyde when he asserted, "A Nation which once was, as everyone admits, one of the most classically learned and cultural nations is now one of the least so . . . the failure of the Irish people in recent times has been largely brought about by the race diverging during this century from the right path."[70]

The causal connection between degeneration and nervous disease at the Salpêtrière was brought to a wider audience by Sigerson's translation of Charcot's *Leçons sur les maladies du système nerveux* into English.[71] However Charcot's theatrical style and methods were not well received in some London or Dublin circles. Sigerson's translation of Charcot's treatise was generally praised by the *Lancet*, but skepticism was expressed in regard to the cases of "aggravated hysteria," some of which had "no counterparts in England."[72]

Sigerson's translation of a second series of the same book was published in 1881 and the notice of this volume in the *Dublin Journal of Medical Science* was unenthusiastic; the reviewer complained that although the book "well repays perusal, it will be seen that the present state of science cannot always be fully represented in its pages."[73] Sigerson's French influences therefore prompted a degree of suspicion in parts of Dublin's scientific fraternity. Charcot's use of the Platonic archetype in his classification of diseases did not sit well with the growing materialist mood in the medical community this side of the Channel.[74] English-speaking observers of Salpêtrière studies were generally wary of the various biophysical approaches favored on the continent. Instead, they tended to account for the data in terms of *psychological* factors such as "expectant attention."[75]

But beyond the biophysical research of Paris and its implications for degeneration, another feature of the nervous system was beginning to draw Sigerson's attention in the 1880s. Earlier, in the 1860s, Paul Broca's research at the Salpêtrière had noted a relationship between right-handedness and left-hemispheric brain specialization for language abilities. Both language and handedness appeared to relate to the separation of functions between two halves of the human brain, also known as lateralization.[76] In tandem with these discoveries, as Gregory Radick observes, was the evolutionary debate around the human–animal border in the late nineteenth century.[77] This discussion raised further questions about the origin of language, and ideas that language evolved gradually from prehuman ancestors began to be investigated. If such a theory could be demonstrated empirically, the implications were that primate communication might offer glimpses of earlier human linguistic capabilities. Building on this, the idea that language evolved not from vocalization but from manual gestures was gaining ground among materialist evolutionists. They favored the idea that gestures survived as a behavioral fossil coupled to speech, and that Homo sapiens prevailed because of a switch from manual to vocal language that allowed them to use their hands for the manufacture of tools and weapons, and their voices for instruction.[78] A prime concern in this area of science, was therefore to investigate the possibility that primates might exhibit signs of lateralization.

Sigerson's publications while back in Dublin in the 1880s focused primarily on the phenomenon of lateralization, and he documented case studies on alternate paralysis from right- or left-sided brain lesions, and on dextral preeminence.[79] However, in his case the subtext was an abiding belief, like Max Müller's, that language arose suddenly and replete with conceptual meaning and was the exclusive domain of humans. It will be necessary here to delve a little deeper into some nineteenth-century empirical spaces of the lateralization debate, to place Sigerson's quest to keep his science on a par with the Romantic worldview espoused by his Literary Society compatriots.

From an evolutionary point of view, for handedness—and by implication language—to have evolved, it had to be a natural variation rooted in structure or physiology, rather than an acquired or learned habit.[80] The evolutionist's challenge was therefore to find the predisposing feature for this phenomenon. Broca had noted brain asymmetries in which there was a correlation between the heavier and more highly developed sides and a preference for hand use on the contrary side.[81] Experiments by David Ferrier in 1873 on electric (faradic) stimulation of the exposed cortex of a dog, cat, rat, and monkey supported this by demonstrating limb and body movements on the contralateral side. Ferrier saw implications here for human motor functions and concluded that the left hemisphere was the driving side.[82] In 1871 William Ogle had championed the view that the blood flow to the hemispheres determined handedness by scrutinizing the various structures and workings of the subclavian and carotid arteries, and as a result of his research affirmed "that the mental faculties concerned in speech are in the great majority of men lodged in the left hemisphere."[83] The same year, the London physician Philip H. Pye Smith proposed that "natural selection" accounted for right-handedness. He opined that left-handed warriors would have been mortally wounded without the protection of the shield.[84] As Harris points out, Darwin may have been aware of many of these studies when he noted briefly in 1875 that handedness was "an inherited variation" and that "left handedness was a departure from the normal assymetrical condition."[85]

In 1884, adhering firmly to a transcendentalist rationale, Sigerson presented a paper to the Royal Irish Academy on his studies of dextral preeminence in which he argued that primeval man was ambidextrous. Here, he systematically refuted the brand of evolutionary argument put forward by Broca, Ogle, Ferrier, Pye, and others, suggesting that such theories were in danger of viewing left-handedness as a "monstrosity," and of falling "within the domain of teratology." Instead, Sigerson concluded, "The simple occupations of the early world . . . suited either hand or engaged both." The influence of French neo-Lamarckism on his postcranial studies was also evident. Right-handedness, he suggested, "was a habit acquired to facilitate tasks of increasing complexity, and is transmitted by inheritance, constituting a tendency or predisposition in the offspring." Contrary to growing medical opinion, he held that it was "highly improbable" that the existing preference for use of the right hand depends on "the imperative dictates of the anatomical structure of the human organism."[86]

While acknowledging the fact that from ancient times special attention was given to the right hand—"at least in certain races"—he worked to demonstrate that this was not a universal phenomenon. Evidence was marshaled from a wide range of sources, from ancient art to his own routine measurements with a dynamometer for measuring grasp strength.[87] Empirical evidence, he noted, had been recorded in the observations of Vasco da Gama on a voyage to Calicut, where the inhabitants of Melinda—"a polished and flourishing people"—were all left-handed. Sigerson also points to the "left handedness of the Hebrew race," as unearthed by the more speculative observations of Dr. Erlenmayer, who, while noting that most "Aryans" write from left to right, and most Semitic peoples write from right to left, "maintains that the writers of the Old Testament and probably the early Talmudists so wrote, because they used their left hands."[88]

In the true spirit of Romanticism, Sigerson was not convinced that language "complex and organic, as it is in its intellectual character as the sign or symbol of an idea," could be localized to "so limited and defined a seat in the brain."[89] He mobilized science where possible to seek evidence of a common human mentality, and at the same time he worked to dispute any evidence of "higher" cerebral connections between man and ape. Sigerson therefore rejected attempts to locate a cerebral center for speech, and by extension, ruled out a propensity for right- or left-handedness among ancient populations. The belief that language and thought were inseparable—so fundamental to revivalist notions of origins—is also threaded through his scientific approaches to physiological inquiry, as we shall see.

In relation to the blood supply arguments of William Ogle, he quoted Erasmus Wilson's alternative findings that when any "lateral inequality" is shown to exist, "it has appeared to favour the right side and not the left," and under such circumstances, "the right hemisphere should have the better blood supply."[90] Sigerson drew on his metalloscopy experiences at the Salpêtrière to refute Ferrier's experiments on motor function, and made the counterargument that those parts of the brain that had been regarded as motor centers have in fact no direct influence on "motion," but are really centers of tactual sensibility. He acknowledged here the work of Charcot in demonstrating the existence of motor centers in human brains, but it was Sigerson's contention that any human–animal links in the area of the nervous system were the "general property of the animal machine" and would be limited only to areas of "tactual sensibility."[91]

The Royal Irish Academy article continued in this vein, disputing in rich detail the "imperative dictates" of lateralization in both the human and the nonhuman organism. The evolutionary arguments put forward—particularly by Ogle—implied, according to Sigerson, that animals with the same vascular arrangement as humans should show similar right-sided propensities due to increased flow to the left hemisphere. Drawing on his broad knowledge of zoology, he asserted that marsupials, rodents, cetaceans, beavers, and the chimpanzee should all share such an affinity, but only monkey tests had been cited. He concluded, "Hence if the statements made as regards the presence of dextral predominance in these animals be correct, this very remarkable conclusion is deduced, namely that dextral predominance exists in them, not on account of structure, but in spite of structure." Sigerson went on to point out that although the chimpanzee had the closest arrangement to man, other monkeys such as macaques were observed by the "evolutionist" researchers and so must have *acquired* the propensity, if it was noted, because there was no physiological evidence to support it. Sigerson does, however, note an "observed" case of "unequal lateral development" relating to a seal in the Dublin Zoological Gardens, but added, "in consequence of which it swam in a circle."[92]

However, despite Sigerson's misgivings, the gestural origins of language theory had been given a major boost in the 1870s by Ogle's contention—based on experiments at the London Zoo—that twenty out of twenty-three monkeys displayed lateralization. He concluded that "monkeys like men are as a general rule right handed."[93] In a letter to Darwin thanking him for his copy of one of the first printings of the *Descent* in 1871, Ogle mentioned his recent findings on monkey-handedness and suggested that they offer further evidence of the kinship between man and ape. These and other works, particularly by Joseph Deniker in Paris, appeared to underpin this view.[94] Sigerson, however, even in the 1880s was maintaining that "it was quite possible that the alleged facts" about right-handedness in monkeys "were insufficiently, and therefore inaccurately, observed." He cited unnamed "contradictory accounts" with which "my own investigations coincide." These were observations on the Sphinx baboon, which had been seen to fling projectiles with both hands, and "the Aye-Aye," which, according to Mr. Bartlett the London Zoo superintendent, "used only the left hand in feeding." Sigerson argued that it would require the most "scrupulous vigilance" to eliminate from such a test the influence "which right handed men may have had upon animals apt to imitate and probably taught more or less consciously by their human attendant or proprietor."[95]

Clearly, Sigerson was convinced that no evidence to support lateralization or the propensity for kinship with monkeys would come to light. His scientific inquiries had convinced him, like Étienne Bonnot de Condillac, that humans share with animals the innate tendency to make certain natural gestures in response to fear, joy, love, fright, and so on. Such gestures constitute a rudimentary mode of expression and communication in these simple situations, but do not constitute the language that is peculiar to man.[96] For his national vision to transpire, primitive peoples had to have been the vessels of revealed religion, whose folklore, later captured in medieval Irish texts, retained elements of that accomplished time. In Sigerson's words, "It was the Heroic Age of Ireland, when, as in Greece and Rome, all was Gigantic, Titanic, or Divine,"[97] and thus, for him, it was crucial to believe that degeneration in this context had occurred. Any research by progressive evolutionists that established a language link between apes and humans would unsettle his reading of "primitive" Irish society and wholly undermine the "scientific thesis" so fundamental to the National Literary Society's sense of Gaelic origins.

Language and the National Vision of Eoin MacNeill

Growing up in the glens of Antrim, Eoin MacNeill (1867–1945) was profoundly influenced by the local Irish-language traditions that reflected his Gaelic ancestry, in an area, which later became a major focus for Ulster-based Gaelic revivalists.[98] He was educated at St. Malachy's College Belfast and was awarded a degree in constitutional history, jurisprudence, and political economy in 1888, having attended law lectures at Trinity College and the Kings Inns, Dublin. In 1887, MacNeill obtained a junior clerkship in the accountant-general's office in the Dublin law courts. He was the first clerk in the office to be appointed by competitive examination rather than patronage and was also the first non-Anglican member. The same year, he embarked on his studies of the Irish language, which, from 1890, also encompassed Old and Middle Irish under the Jesuit scholar Edmund Hogan. These pursuits reawakened an interest in both Irish history and spoken Irish, which prompted annual visits to the Aran Islands from 1891 to 1908. As a result of his research, MacNeill developed a theory of Irish identity that stressed cultural factors—particularly language—over state power.[99] Throughout the 1890s, he contributed articles on language to the *Irish Ecclesiastical Record* and the *Gaelic Journal* and took a leading role in the founding of the Gaelic League under the inspiration of Douglas Hyde's 1893 address "The Necessity of De-Anglicising Ireland."[100] From 1893 to 1897 MacNeill acted as secretary to the Gaelic League and edited the *Gaelic Journal* from 1894 to 1897.

Like Sigerson, MacNeill's vision was certainly a Romantic one, in his belief that the true basis of the Irish Nation was to be found in the remote Gaelic past and that the language was the lifeline of nationality. According to MacCartney, MacNeill's brand of cultural nationalism was almost a copybook reproduction of that preached by the great European Romantic Nationalists such as Herder. Elsewhere, such as in Louis Le Roux's biography of Patrick Pearse, MacNeill is depicted as a type "keen on the study of metaphysics" and "de-

FIGURE 18. Eoin MacNeill (1867–1945), civil servant, historian and founder of the Gaelic League. Reproduced by kind permission of UCD Archives UCDA LA30/PH/325.

voted to the study of Plato and Montesquieu."[101] As a youth, MacNeill would have been introduced to the Scottish Gaelic tradition of Ossian as evoked by Macpherson in his controversial eighteenth-century translations.[102] Later in his *Phases of Irish History*, MacNeill commended the work of Scottish scholars, namely, George Buchanan and Macpherson, for rediscovering and bringing the Celts to a wide European audience. As he put it: "The arrival of James Macpherson marks a great moment in the history of Celtic literature. It was the signal for a general resurrection," and he dismissed the rumpus that accompanied the publication, adding "not withstanding the controversy that soon arose about the authenticity of Macpherson's compositions, their direct influence and vogue went on increasing for half a century."

But MacNeill had good reason for acknowledging a debt to Macpherson. He believed that the whole technique of the poetry of modern nations was Celtic in origin, and extolled in the *Gaelic Journal* the "perfection" of the rhyming stanza and "the beauty of early Irish poetry," which was "as distinct from ordinary language as the heavens are from the earth." The same piece asserted that "in ancient Gaelic literature was to be found the purest manifestation of the Irish mind."[103] Traditional accounts of MacNeill as a conservative Catholic Romantic Nationalist certainly fed on such gushing pronouncements, but to focus solely on this Romantic *rhetoric* is to miss the regenerative dimensions of his language project. A more detailed consideration of MacNeill's linguistic spaces reveals that the vehicle that carried his vision was none other than a form of Gaelic "modernism."

Attempts to define the nation and subsequently use this definition to support ideal forms of social and economic organization were contested on many levels in late nineteenth-century Dublin. Nationalists of an Anglo-Irish, Gaelic Irish, and Socialist stripe all trawled the archives of history, mythology, and folklore to support calls for landownership systems based on cooperation and communalism as expressions of national identity and as a means of regenerating agriculture, rural life, and the idea of nationhood. Although many, like Sigerson, were captivated by the cooperative potential of Irish agriculture and looked to the existence of a preconquest Gaelic commonwealth—a form of Celtic communism—to establish a native pedigree for their cooperative ideals,[104] MacNeill was not convinced. Certainly, nationality for him was a question neither of race nor government. As he put it: "A nation is a species of the genus civilisation, a state is the species of the genus Government." In a similar vein, Sigerson had extolled the virtues of a primitive Celtic nobility, foregrounding systems of self-sufficient communities over state control. To MacNeill, however, the nation/race debate was exacerbated by phrases such as "the Celtic race," which he claimed were "all pure figment, serving no purpose except the rekindling of insane animosities," and he added, "the less civilised men were, and the nearer they were to the brute, the more would they exalt racial antagonisms."[105] This disengagement between race and nationality had surfaced elsewhere. Arthur Balfour, for example, in *Nationality and Home Rule*, had argued that culture and tradition, and not biological race, were the glue that held nations together.[106] Although his anti–Home Rule argument that the Irish people did not have a "national" identity distinct from Britain may have been jointly queried by both Sigerson and MacNeill, Balfour's contention that Irish culture was not suited to modernity—although inconsequential to a Literary Society Nationalist—reflected the *very issue* that the founders of the Gaelic League set out to address in the 1890s.

Although the intellectual traditions behind their respective language visions were grounded in Romanticism, MacNeill's particular brand unfolded and was realized within an empiricist mechanistic schema in contrast to the idealist and organicist nature of Sigerson's. The language revival was therefore apprehended locally on two epistemological fronts. Ironically, despite his Romantic Nationalist milieu, MacNeill's *scientific* approaches to linguistics sought out the precision of a uniformitarian methodology over the speculations of Romantic science. Unlike Sigerson's leanings toward notions of "a vital fusing force which attracted all newcomers into the nation," the development of the "individual" Irish personality was central to the philosophy of MacNeill's Gaelic League idea.[107] If we accept that the raison d'être of the Gaelic League was twofold—to establish a new nation and to define it by the use of the Irish language—then material concerns were inextricably linked to cultural ones. The Gaelic League from the beginning linked the language revival to Irish economic regeneration.

MacNeill's approach is evident first in his dispute with Henry Maine, the British jurist and legal historian. Maine's analysis of the Brehon Laws (early Irish Law Tracts) led him to interpret the conquest of Ireland as the coming together of two societies at different stages of social evolution, with the more advanced wrongly assuming that the less advanced society was irredeemably other.[108] Maine drew parallels with "Native law" in India to underpin claims that the Irish were a primitive people. MacNeill's criticisms of such an interpretation and his claims that Maine had "come to Irish Law as a happy hunting ground for primitive big game" have again placed him in the nonprogressive evolutionary Nationalist camp.[109] However, in her critique of this altercation, Heather Laird points out that MacNeill's reason for rejecting the Irish/India parallels was his own interpretive focus on what he pinpointed as the "modern" aspects of the Brehon Laws.[110] In 1852 the government appointed a group of commissioners to translate, transcribe, and publish the ancient laws of Ireland. In the official nineteenth-century translations of the laws, on which Maine had drawn his theories, early Irish society was based on common descent and communal ownership.

However, MacNeill's own studies had elucidated a sophisticated contractual model of early Irish society, which was based on familial units but not structured in the way interpreted by Maine and others. In *Early Irish Laws and Institutions*, a compilation of speeches and lectures regarding the construction and execution of laws in early Ireland from the Pax Romana to the Brehon Laws, MacNeill succinctly defended the idea of Ireland as a cohesive nation, not a conglomerate of different tribes and clans.[111] Drawing on his philological research, he proposed in *Early Irish Laws and Institutions* that the term "tuath" meant territorial or political unit ruled over by a king, the land of which was held in severalty by various joint families of different descent. Far from a form of primitive communism, this was more an extreme form of private property. Thus, MacNeill constantly rejected any trace of a clan system and dismissed the terms "tribe" and "tribalism" in future publications. He denounced scholars who spoke of Irish tribal society as "primitivists" who sought to degrade the pattern of life in ancient Ireland to "the level of the Australian and central African aborigine."[112]

For MacNeill, the term "tribalism" implied a complete negation of national unity. He became convinced that there had been a general misreading of Irish history over the past century due to incorrect interpretations of Irish traditional laws and institutions, resulting from poor translations. His own interpretations proposed that Irish law was a national system, and he asserted that the laws did not differ from region to region but remained consistent and nationally applicable in all parts of the island. His interpretation of early Irish law justified for him the view that the Irish had once formed a unified society. But not content with the pervasive Romantic nostalgia engendered by "having reconquered the past for the disinherited" as was the case with the Literary Society Nationalists, he called on Irish scholars to "proceed without delay to conquer the present and the future."[113]

These calls coincided with the wave of Nationalist fervor sweeping through Europe in the late nineteenth century, and as each new nation struggled to demonstrate its "superior fitness" a strong central government became paramount to that success. The contemporary social Darwinian model, as Herbert Spencer had earlier advanced, saw the development of society as an evolutionary progression from simple undifferentiated homogeneity to complex differentiated heterogeneity. According to Spencer, movement from the simpler militant state (structured around relationships of hierarchy and obedience) to the more complex industrial state (based on voluntary contractually assumed social obligations) happened according to the universal law of evolution. As Peter Bowler observes, this intensification of national rivalries in late nineteenth-century Europe allowed the application of the social Darwinian model to emerging societal frameworks.[114] MacNeill, on the other hand, redefined these models to Gaelic League specifications, and replaced the evolutionary scheme with a causal uniformitarian one. By calling on the authority of scientific objectivity, he could make the assertion that the nation existed as a true "cultural" entity from primitive to modern times.

MacNeill posited that whereas contemporary and ancient Empires were states held together by the power of a central ruler, the forces of union that underlay the Gaelic idea were embodied in the language, literature, law, traditions, and religions of the people. Claims for Irish political autonomy, he wrote, "have never been based on Ireland's having existed in former times as a state, they were based on the existence of an Irish Nation throughout the ages of Irish history."[115] For MacNeill, therefore, the idea of a cultural revolution was much more important than any political change in Ireland's status, and contemporary European linguistic thought would be implicit in such a change, especially linguistic theories with a scientific underpinning. The specific spaces of the Gaelic League, which were to enable MacNeill's commitment to science, vividly illustrate this point.

The Gaelic League was founded in 1893 with the merging of the Society for the Preservation of the Irish Language and the Gaelic Union, and its objectives resonated with sentiments similar to other European-language movements. The Gaelic League emerged from a European-wide tradition of antiquarian scholarly research and was led by an intellectual elite who officially professed that it was both a nonsectarian and nonpolitical organization. However, from its inception in 1893, the Gaelic League disregarded the basic assumption of the previous societies—that Irish was a dying language that needed to be documented before its passing—and insisted that the language could be revived as a means of mass communication in Ireland. Its aims were the preservation of Irish as the national language, its propagation as a spoken tongue, and the cultivation of Gaelic letters.[116] Not only did the league place an emphasis on the spoken language for the first time, it also fostered and encouraged a "modern" Gaelic literature.

Like the theater movement the Gaelic League grew out of the rise in opposition to imported English culture during the 1890s. As a result, one of its main objectives became the reinvention of the Irish language and its intellectual inheritance as the cornerstone of Irish identity. Small branches were set up all over the country to provide informal instruction outside the official education system, which was hostile to Irish-language studies. Many of the branches were set up by voluntary organizers who traveled the country to coordinate classes and activities. Their work was analogous to the efforts of cooperative enthusiasts who set up the new Irish Agricultural Society branches.[117]

The central branch or Ard Craobh, which was set up in Dublin City, recorded 440 members by 1899. Here too, as Timothy McMahon tells us, the zeal of some Gaelic enthusiasts drove the "merely curious" away.[118] The satellite groups set up in the Dublin suburbs were distinguished by their class diversity. They emerged both in affluent townships where "the respectable white collared, trim suited suburban Gaelic Leaguers"—as O'Casey described them—"liv[ed] rosily in Whitehall, Drumcondra, Rathgar, Donnybrook and all the other nicer habitations of the city,"[119] and by the Great Southern and Western Railway works in the Liberties, Blackhall Place, and Inchicore, the heart of working-class Dublin. Yet other branches appeared in the "squalid and declining areas" bounded by the Royal and Grand Canals. In tandem with these class divisions of the Gaelic League, ran an occupation-specific theme within its urban territories. Craobhacha emerged within diverse gatherings such as the Wellington Quay working men's club, the Metropolitan Housepainters' Union, and the Antiquarian Cyclist Society of Ireland.[120]

Embedded in the Gaelic study classes undertaken by each branch, was the message that language was a fundamental component of not just cultural but also economic revival. As MacNeill stated, "The material side of life cannot be ignored. If we ignore our corporeal wants we are not likely to keep up high thinking for very long." And he continued, "The movement must be radical. It must begin at the beginning of every branch of secular knowledge, and go rapidly ahead until it covers the entire ground."[121] The first edition of the Gaelic League's newspaper *An Claidheamh Soluis* in 1899 declared that it was of utmost importance to link socioeconomic and linguistic regeneration: "The decline of Irish is a symptom of the greater socio-economic failure of the late 19th century and secondly the revival of Irish is itself a cause of development. The language, the industries and the very existence of a people are all interdependent, and whoever has a living care for the one cannot be unmindful of the others."[122] By the turn of the century, the executive committee of the Gaelic League was calling on all its members "to give active support to industries in Irish speaking areas." Another member of the league, the journalist and polemicist D. P. Moran, founder of the Nationalist newspaper the *Leader*, presented his views on cultural identity, language, and development, in an essay titled "The Battle of Two Civilisations," in which a link between language and material progress is again made.[123] The Dublin Unionist MP, Horace Plunkett, included in his publication *Ireland in the New Century* a chapter on the Agricultural Co-operative Movement, of which he was a founder, and its important links with the Gaelic League. He made the claim that "the League's aims of consolidating the Irish language, formed a valid part of the process of socio-economic regeneration in Ireland."[124]

The Gaelic League's vision of the Irish language, therefore, was not merely a cultural artifact for study by antiquarians. The emphasis generated by

MacNeill on culture and language as a resource was in opposition to dominant contemporary theories of economics. The Gaelic League raised questions about the dominant models of development coming to them from the English-speaking world, and they attempted to challenge them, as this quote from Moran's essay suggests: "At first blush it may appear a far-fetched idea that there is a strong connection between the development of a native civilisation having its roots in the native language and the production of economic wealth. We have now come to see that land, though an indispensible, is by no means the main source of modern economic wealth. Human skill in all its manifold manifestations has taken the premier place."[125] So, far from being a nostalgic revival movement, the Gaelic League actively concerned itself with the material problems of a new Irish nation and recognized a link between the language revival and the regeneration of the Irish economy. Its journals drummed up support for "native manufacture," advanced the argument that there was a link between the decline of the language and the loss of industry in rural Ireland in the nineteenth century, and it instituted its own industrial committees.

The Gaelic League saw itself as a "modernizing" and progressive force within Irish society during the revival, yet its program for development was informed by a desire to follow a path to modernization that differed from the English-speaking world. As Mathews points out, "Colonial modes of development invariably configured tradition as a direct antithesis to modernity."[126] But in contrast, the league deployed the past to challenge the present. It is thus possible to view the work of the Gaelic League as presenting an alternative modernization process that was informed by a belief in the neoteric potential of precolonial and early Christian Irish cultural forms. MacNeill saw in early Christian texts a material vocabulary that placed the Gaelic language on a par with Latin and Arab contributions to the intellectual life of early medieval Western Europe. Surviving copies of manuscripts such as *An Irish Astronomical Tract* bore witness to its particular use for educational purposes in a broad sweep of secular sciences such as physics, chemistry, geology, physiology, botany, and human geography, and provided the archives with what had become an obsolete Middle Irish technical vocabulary.[127] Rather than looking to the past for Romantic inspiration, MacNeill sought to mine the dormant resources of the Gaelic past to kick-start the Gaelic present. As he put it, "The world's material progress is not like the motion of a detached projectile, but rather like the motion of an endless belt or chain . . . you have not to overtake anything that is past and gone, you have only to join in with the movement at the point where it is at present passing you." He saw it as his duty, therefore, to remove "the material handicap from the back of the national language," and begin a program to once more "bring the Irish language abreast of the material needs of the nation."[128] To do this he turned to contemporary science.

As we have seen, MacNeill's Gaelic League spaces revolved around the twin aspirations of socioeconomic and linguistic regeneration. This partnership would ensure that the scientific mechanisms of material revival would diffuse readily into the linguistic arena. But MacNeill's approach to science in his language revival project had been forged not only by the spaces of his economic endeavors but also by his legal spaces. Elements of Stephen Alter's observations on conceptual transfer in the nineteenth century, as mentioned earlier, have also been explored by Lorraine Daston and Michael Stolleis in their critique of the relations between the "laws of nature" in natural philosophy and "natural law" in jurisprudence. The authors suggest a common conceptual matrix that made the analogy between legal and natural orders compelling.[129] In this way, it becomes clearer how MacNeill's training in jurisprudence had already preempted his methodological approach to language studies.[130] And in terms of situating that methodological approach, MacNeill's unpublished observations on the laws of nature indicate a preference for early nineteenth-century causal uniformitarianism over Sigerson's organistic post–*Origin* biology:

> The notion of the laws of nature has had great value for the development of physical science, it has enabled scientists to classify phenomena that once appeared at random and detached from each other. It has enabled them to anticipate, to discover and to invent, to rely on the existence of data that had not been observed, to reach out into the unknown, and above all to disclose in the universe the presence of order even more wonderful than any ancient seers and poets thought to have perceived. It is another thing altogether, to inculcate, in the receptive minds of half reasoning children, the notion of a self-existing Law of Nature.[131]

The notion that valid causes of unexplained events could be inferred by analogy with forces observed operating in actual experience had been explored in the early nineteenth century by scientists such as the geologist Charles Lyell and the astronomer William Herschel. Moreover, Herschel had urged that the method be applied to the study of human institutions such as language. William Whewell's *History of the Inductive Sciences* had also proposed the notion of a parallelism between the natural sciences and language study. Both disciplines, he suggested, sought to ascertain "a past state of things by aid of the evidence of the present."[132] MacNeill alludes to these earlier nineteenth-century causal uniformitarians when he remarks: "Great scientific thinkers and observers have been able to show that many phenomena exhibit a common tendency of matter, they have inferred that other things will happen accordingly and have been able to verify the influence. Even the unscientific arrange acts continually on the conviction that the things which have come to pass before, under certain circumstances, will come to pass again and again under the same circumstance . . . in other words nature implies a fixed order in the physical world."[133] Combining this aspect of nineteenth-century science with the material influences of the Gaelic League, the scene was set for MacNeill to align his linguistic vision with new developments in European comparative philology. A project as immense as his language regeneration mission would require a widespread understanding of the rudiments of Gaelic, and therefore the systematic dissemination of its inherent regulatory laws would have to be undertaken.

To achieve this, it was necessary for MacNeill to follow in the linguistic tradition first proposed by Jacob Grimm in *Deutsche Grammatik*, in which Grimm used phonetic correspondences to reveal the systematic "rules" that bound together the Graeco-Latin and Germanic languages in a practice that became known as "Grimm's law."[134] By focusing on "living language" and speech in this way, the project opened up further possibilities for MacNeill in another arena. Selecting linguistic mechanisms relating to phonetics rather than word structure enabled him to explore constructions of national identity.

The linguistic anthropologist Brigittine French has examined the role of linguistic science in MacNeill's vision of Irish national identity and scrutinizes the links between "scientific" linguistic scholarship and national identity formation embedded in his grammatical analyses of Irish, particularly sound systems. She shows how MacNeill invoked a scientific framework in his representations of Irish sounds to support his call for an autonomous Irish people, "an ethno linguistic community, conceptualized as inherently distinct from the English language and British people and naturalized by 'objective claims' to knowledge."[135] Judith Irvine and Susan Gal have also outlined the semiotic processes by which linguistic difference becomes mapped onto social difference in this way. They describe how the regimentation of phonetics enables the development of standardized orthographies and, in turn, standardized orthographies facilitate the proliferation of textual materials for vernacular literacy. This process often becomes key to imagining an independent national community.[136] So, from a Nationalist point of view, this particular contemporary scientific methodology was part of an enabling late nineteenth-century discourse from which to sow the seeds of a Gaelic revival.

This linguistic appeal to science over history places MacNeill well within the intellectual ferment of the neogrammarian movement. As Olga Amsterdamska points out, the movement was defined by its assertion of a particularly scientific and historicist philology.[137] The neogrammarians sought to make historical linguistics an exact science in line with those natural sciences that had made advances in the nineteenth century and that held strongly to the universality of natural laws, realistically conceived. For the proponents of the new movement, language was not a supraindividual organic identity with its own growth and life, as von Humboldt, August Schliecher, and later Ferdinand de Saussure asserted. It simply had its existence in the individuals composing a speech community, and linguistic changes were changes in individual's speech habits. The neogrammarians viewed these sound changes as mechanical processes that take place according to laws that admit no exceptions.[138] As the prominent neogrammarian S. A. Leisken stated, "If one admits optional, contingent and unconnected changes, one is basically stating that the object of one's research, language, is not amenable to scientific recognition.[139]

In sum, the neogrammarians rejected historicist beliefs about linguistic growth and decay and the Romantic conception of language as an organism; instead they set out to formulate a theory of linguistic change based on a definition of language as a psychophysical activity and on the assumption of causal uniformitarianism (the belief that language is subject to the same laws in all periods of its development). MacNeill's aim to give phonetics a more scientific direction through his analysis of Irish sounds situates him well within the new generation of linguistic scholars who took their direction from Hermann Osthoff and Karl Brugmann's 1878 publication *The Morphological Investigation in the Sphere of the Indo-European Languages*, and saw themselves as "linguistic scientists" rather than philologists.

In keeping with the neogrammarian's dislike of historicist speculation around periods of growth and decline, it is evident from a critical examination of MacNeill's scholarship that he downplayed the importance of any historical continuity in orthographic representations of the Irish language and tended to chastise scholars who focused on the oldest forms as the inherently most preferable ones: "The student of Irish is often tempted to stand up for some isolated usage, even when it survives in some very small district, when he finds authority for such usage in the older written forms of the language."[140] The neogrammarian manifesto held that there was no difference between ancient and modern periods as far as linguistic changes were concerned. The spoken dialects of Europe had been a focus of linguistic attention since the Romantic movement sanctified everything connected with "the people," but this new movement made contemporary everyday speech a vital field for scientific investigation because of the light it could shed on linguistic change.

As neogrammarian investigators were called on "to pay attention to speakers, not just abstract language, to understand language as an activity, to focus on language use, and to look at the spoken language of both speech community and individual," MacNeill was instructing his colleagues that "a living language must be learned through the ear" and advising learners of the language to live among native speakers: "The only thing we have to build on is the Irish we find spoken . . . instead then, of turning away in distaste from differences of dialect, we ought as far as possible to turn to them, and master them."[141] He also questioned the use of poetry and folklore for learning spoken Irish and claimed that texts for learning Irish were almost invariably selected on account of their celebrity rather than their practical usefulness, praising instead O'Leary's short book on the vocabulary of the bicycle as "a small but significant" move in the right direction."[142] This great mistake had done immeasurable harm to the study of Irish, tending to create a gulf between the modern spoken language and the printed literature: "If Irish is to become once more the language of a nation, it must shape itself to express all the thoughts of a modern nation's life."[143]

The neogrammarian focus on "living" speech and the manner in which it was transcribed became a major preoccupation for MacNeill throughout the 1890s. It was at this time, by means of a series of submissions to the *Gaelic Journal*, a monthly publication of the Gaelic League, that the political potential of neogrammarian linguistic science unfolded for his readers. In these articles, MacNeill stressed the inherent differences between English and Irish sounds and their orthographic expression. He did this by resolutely establishing isomorphic relationships between individual sounds (phonemes) and their written representations (graphemes) based on scientific principles. The orthographic conventions that he proposed had as their aim a systematic reduction of graphemes that would give Irish orthography a more accurate phonetic representation. This is amply illustrated in his discussion of the verb "bi": "Hence it is clear that 'biodh' is the correct spelling, and in careful examination of MSS of the period when the modern orthography of Irish first took shape, no other spelling is found. In every case the first of the two spellings is the more correct phonetically and has the additional advantage of being much the shorter. In each instance besides being phonetically more correct we save the writing of several silent letters."[144] This distinct phoneme to grapheme representational system would distinguish the Irish language from others and, as French has pointed out, the Irish representation of sounds could stand in opposition to English ones.[145] MacNeill made no secret of his disproval of the English phonetic and graphemic systems and their effective co-optation of the Irish language:

> Owing to the fact that most of us learn to spell in English first, many imagine that spelling according to the English system, or want of system, has a kind of divine

> right or prior claim to the representation of sounds. They imagine that "sh" is a more natural way of representing the "slender" sound of "s" than the Irish way, that "v" is the true way of writing the sound of "b," and so with other characters. And they call these devices "phonetic." Written phonetically had come to mean spelled in the English way. It is time we should hear no more of it. "Phonetic" means according to sound, and the spelling bíΓ is far more scientific and phonetic than "veesh."[146]

In another earlier essay on Irish spelling conventions in the *Gaelic Journal*, he criticized the popular notion of the term "phonetics," which he regarded as doing nothing other than assimilating Irish sounds to English writing. Parodying the political aphorism "One man one vote," MacNeill's perfect Irish alphabet was encapsulated in the motto "One sound one symbol,"[147] and he consistently called for a more scientifically precise definition of phonetics: "A further complexity has been introduced by the unscientific theorizing of late grammarians, taking third person singular as stem, these short forms have everything on their side except the partial vogue given to their unwieldy upstart rivals by grammarian pedantry within the present century. If the X Rays of the linguistic surgeon show the dh to be a mere excrescence on the bone, an excrescence that disfigures appearance and impedes motion, surely the pain of the trifling surgical operation necessary to get rid of it is worth enduring."[148] Directly antithetical to Sigerson's organicist, all-encompassing approach to the language revival, MacNeill tended toward the isolation and control of linguistic mechanisms in his work, essentially separating the language from its environment. But overwhelmingly, his strategy for establishing a modern Gaelic-speaking nation can be seen as an appropriation of European neogrammarian linguistic science and its empiricist methodology. His scholarly recommendations for spelling in Irish sought to advance a national mindset rather than a colonial one. From this linguistic perspective, an independent Irish writing system with distinct phoneme-to-grapheme representation came to stand for an independent nation warranted entirely by claims to science.

By targeting the interplay between linguistics and science, this chapter has worked to open up new lines of inquiry into the ways in which the origins of language were comprehended in late Victorian Dublin. The episode has endeavored to bring two arguments to the fore. First, it contends that the course of the Gaelic revival in Dublin was charted, in a major way, by scientific discourse emanating from Europe. And second, it proposes that the nature of Nationalist scholars' engagement with this science was profoundly influenced by their Dublin spaces.

The focus on the reception and mobilization of scientific ideas by Sigerson and MacNeill during the Irish language revival has brought to light something of the manner in which Irish fin-de-siècle cultural, social, and political spaces have enabled and constrained the interpretation and dissemination of scientific meaning in Nationalist circles. The discussion has probed the ways in which the intellectual spaces of Sigerson's National Literary Society and MacNeill's Gaelic League discouraged particular strands of scientific thought and encouraged others, and demonstrates how Sigerson allowed Müller's idea of "the science of language" to germinate, whereas MacNeill provided fertile ground for a neogrammarian "linguistic science" to take hold. It has also drawn attention to the import of both a "transcendental" and "empirical" dimension to the revival, and to the ways these factors facilitated an appeal to scientific authority on two distinct fronts.

George Sigerson envisaged the new society as an organism and a composite of many races, peoples, and creeds and hence rejected the social Darwinian stress on competition evident in progressive ideologies and advocated "mutual aid" and "co-operation." But, as a result, he commended the monogenism and ethical implications of Darwin's theory and used it where possible to underline his belief in organic solidarity. In contrast, we see that in *Early Irish Laws and Institutions* MacNeill alluded to an extreme form of private property, far from Sigerson's form of primitive communism, and rejected any trace of "tribalism." For MacNeill a new Ireland could only thrive if envisaged as a mechanism composed of individuals and governed perpetually by universal laws "like the motion of an endless belt or chain." The goal of his linguistics was the description of the language as a total system, as a necessary prerequisite for the material regeneration of the country. Much of MacNeill's scholarship was devoted to purging contemporary Irish orthography of unnecessary graphemes, preferring "one sound" for "one symbol" and excising "excrescences" from the bones of the Irish language.[149]

Although "Celtologists" like Eoin MacNeill insisted that episodes in early Irish history should be properly translated so that a full picture of the primitive world might be *reconstructed*, the view coming from the Literary Society was that these same episodes should be properly re-created so that the full primitive imagination might be *experienced*. The Literary Society saw the Gaelic League's need to fuel nationalism with historic fact as a betrayal of the primitive "imaginative energies." It held that the efforts of the Gaelic League to convince groups of enthusiasts of factual importance in early Irish writings overlooked the importance of the inherent fantasy and drama. The Literary Society and the Gaelic League therefore shared an allegiance to the ancient texts for diverse reasons: one for the sake of accuracy and the other as a source of inspiration. In conclusion then, it can be said that nineteenth-century scientific theories, whether Darwinian, Lamarckian, or inspired by the earlier Uniformitarian concept that "the present is the key to the past," were mobilized selectively in different Nationalist spaces in Dublin to underpin particular linguistic ideologies and advance different national visions.

Locating Origins in Dublin

In an effort to better understand the production, circulation, and reception of ideas about "origins" in late nineteenth-century Dublin, this book has argued that *siting* the scientific work of a group of scholars within the city and defining and examining the spaces of their professional activities deepens and enriches our understanding of why they held certain convictions and why they took particular stances.

Dublin in this study has worked both as a bounded physical entity *and* as a theoretical device. As the former, however, it did not act as a mere backdrop or container for exploring scientific ideas in a national context—as in "styles of Dublin thought"—nor was it reified as a geographical frame of reference. Instead, the city was called on in numerous spatial manifestations to reveal the more intricate connections between lives and science. As well as uncovering the city itself as an important arena for addressing the geographies of nineteenth-century science, we have also seen that behind the seemingly solid walls of scientific institutions such as Trinity College, the Royal College of Science, and the Anthropometric Laboratory were malleable spaces and hidden intellectual dimensions. These have allowed us not merely to add new territory to an old map, but to transform the entire concept of an urban scientific landscape.

Reflections on hidden spatial dimensions are not new of course. In the nineteenth century the notion was explored by Edwin Abbott in *Flatland: A Romance of Many Dimensions* (1874).[1] This was a fictional portrayal of a two-dimensional world trying to deal with the idea of three-dimensional figures. With *Geographies of City Science*, perhaps we too, like Abbott's characters, can expand our imaginations beyond the limits of our "respective dimensional prejudices," and harness new spatial dimensions to better understand the world around us. The theoretical physicist Lisa Randall, has, for instance, postulated the idea that the world in which we live could be a three-dimensional spatial region within a system of "warped" extra dimensions.[2] But this has a simpler analogy here. Our minds engage with multidimensional activities every day, we organize our thoughts across many dimensions, various parameters crisscross and interweave in ways that are challenging to re-create in our exterior three-dimensional world. David Livingstone reminds us that "making sense of even the simplest gestures and behaviours requires an understanding of the 'imaginative universe' in which the occupants of any particular locality dwell."[3] Likewise, just as our Dublin sites can be pinpointed in space by three Cartesian coordinates, they can also be apprehended by numerus other spatial dimensions that are invisible to the naked eye.

The more abstract territory explored in this book has therefore allowed a new urban scientific map to emerge. Rather than calling on traditional biography, the analysis took account of the notion of habitus and examined the individual dispositions of the protagonists at given junctures in space and time. Political alignments, religious denomination, traditional mores, professional allegiance, and everyday habits directed discussions toward the manifold "spaces" of eight scientific lives as a means to better understanding how science took its shape in the city. This new map of Dublin is demarcated not by a host of Cartesian coordinates, but guided by, as Richard Dennis describes, the "diurnal rhythms" of a life in the city.[4]

Beyond "Ascendancy" and "Nationalist" Science

As Steven Shapin insists, scientific knowledge is made by "locally varying" modes of cultural interaction.[5] This notion, as we noted, problematized traditional approaches to the history of science in the city. For one thing, it profoundly troubled the monolithic agency of religious denomination invoked by previous studies, which invariably conceived Dublin as a Protestant stronghold or a center of Catholic culture into which a universal scientific enterprise found different expression. By adopting a spatial approach, the investigatory parameters of "origins" inquiry in the book were broadened out considerably.

Our travels confirm that Dublin scientific institutions and societies were more than merely "Ascendancy" or "Nationalist" strongholds. Instead, they consisted of multifaceted spatial zones where origins discussion—in areas from physics to botany and from anatomy to linguistics—was both enabled and constrained by local issues. Exploring the negotiation of these Victorian scientific debates from such a perspective has helped to elucidate in finer detail the fin-de-siècle scientific panorama of the city and has drawn attention away from previous historiographical accounts of science in Dublin that *center* on a bifurcated Catholic/Protestant model.

Chapter 1 revealed that the "cognitive exclusivity" of natural philosophical debate in Dublin was not in concert with a united Protestant perspective. In fact, the very notion of "Protestant" science was unsettled by the physicists George Francis Fitzgerald and William Fletcher Barrett, whose contemplations on the ether were firmly domiciled in their Dublin habitus. Similarly, chapter 4 retreated from traditional historiographical presentations of a Dublin Protestant-led racial experiment on Catholic "primitives" in the west of Ireland. The story here also revealed that the local circumstances of both scientists' lives and work inflected to a much greater degree on their engagement with the Dublin site of anthropometric experiment than any pan-Protestant affinities. The cogitations of biological debate and its interplay with Presbyterian mores was also thrown open to spatial scrutiny, and in chapter 3, we saw that physical and intellectual siting again trumped any theological concordance in matters of science. The book also took issue with historiographical presumptions around science and Catholicism in Dublin. In chapter 4, the habitus of both George Sigerson and Eoin MacNeill were found to have directed their science and their vision of a new Irish society to a far greater extent than religious beliefs alone. All the case studies queried traditional perceptions of Dublin's history, which had polarized science into discrete Protestant and Catholic enterprises. But in addition, by employing the concept of habitus and assigning "Dublin's" role to a series of sociospatial cityscapes, less overt spaces of origins discourse have risen to the surface, and new metropolitan cartographies can now be explored.

New Metropolitan Cartographies

This more spatially sensitive approach has allowed for a considerable reshaping of the contours of origin debates in the late nineteenth-century city. For instance, under the advocacy of the new Anthropometric Laboratory, as chapter 3 makes clear, both Alfred Cort Haddon and Daniel John Cunningham amassed empirical data to furnish their own particular plans for anthropology. When issues such as personal goals and challenges to cultural identity were allied to general fears of population decline, racial extinction, and political uncertainty, what really sustained the motivation behind the scientific work of these laboratory directors came into sharper focus. Likewise, Fitzgerald's reluctance to accept the demise of the ether was the manifestation of a specific Dublin habitus in which an accommodation between modern science and Trinity College Unionism was sought. Dublin's many other institutional spaces were also far from being "neutral" containers for science. For the scientists working within the walls of these buildings, engagement with scientific theories was, as we have seen, either facilitated or forestalled by local particularities. The geographical approach also revealed something of the role of nonscientific professional organizations in influencing scientific endeavor in the city. Chapter 4 outlined, for example, that the spaces of both the National Literary Society and the Gaelic League galvanized particular strands of scientific thought in the minds of their members, and deterred others. The more pragmatic ethos of the league enabled a neogrammarian style "linguistic science" to permeate its Craobhacha, but the literary savants chiefly embraced Max Müller's Kantian transcendentalism.

But institutional and professional spaces do not circumscribe the bounds of geographical understandings of science in the city. The scientists in this book also engaged with many other more abstract city spaces. *Geographies of City Science* traversed this terrain by examining the interplay of morphological interpretations with the light and dark city spaces of Moore and Macalister. It found that ideas of form were made to work with and against Darwin's theory in different ways, depending on the local city particularities of a botanist's life and an anatomist's life. Alignments of science with specific city locations were discovered to be spatially important in other ways too. For instance, in chapter 1, we saw how the routines of city life itself and its arc of institutional authority inflected differently on Fitzgerald's life at the center compared with Barrett's life on the city margins. These few examples highlight just how our understanding of nineteenth-century "origins" issues in Dublin was shaped by the coconstruction of scientific habitus and urban space.

In sum, the geographical premise that "science's practices and forms of communicative actions are always grounded in particular settings" has been amply demonstrated in this volume.[6] We have seen how particularities of site, institutional organization, and city location deeply inflected the ways in which the debates around ether, form, humankind, and language were negotiated and interpreted in Dublin.

These concluding pages now outline further ways in which a geography of science in Dublin might be cultivated. The investigations congregate around the concepts of "perceptions of place" and "circuitries of city knowledge," and will draw on examples from the book to illustrate more general points about the wider implications of the study.

Perceptions of Place

The introduction referred to the idea that the scientists' different perceptions of Dublin had a bearing on the way science was differently understood and mobilized in the city. The ensuing chapters have underscored that "Dublin" did indeed mean many different things to the scientists under scrutiny. We are reminded by the individual stories that local encounters with science were shaped by the way practitioners "conceived of their local settings, by what they took their situation to be, as much as by the ontological realities of those venues."[7] As we have seen, the *centrality* of Dublin to Fitzgerald's life and work was also the driving force behind his science, and it was tied in with an attempt to underpin Unionist spaces at a time of precarious political geography. By comparison, the views of his colleague and fellow physicist, William Barrett, were at odds with this sense of Dublin, his perception of "Dublin" opened up new spaces in which the promulgation of spiritualistic theories complemented his scientific research.

Although notions of "Dublin" played a positive role in the lives and science of Fitzgerald, Sigerson, McNeill, and David Moore, for example, they came to have a negative significance for Haddon, whose mind after 1893 was firmly in the Solomon Sea with the Torres Straits Islanders. Haddon's life and science was to feel increasingly "out of place" in Dublin. As the 1890s progressed, he began to feel more isolated in the city "finding not much scope there for my new interests."[8] In a later letter to his friend James Frazer, the author of the *Golden Bough*, he admitted, "The truth is that my being in Dublin was a great hindrance to me, but I have to earn in Dublin the bread that enables me to work at Anthropology in Cambridge for nothing."[9] When the post he was expecting to fill at Cambridge fell through, Haddon's wife Fanny encouraged him to apply for a chair in Aberdeen because it would mean "not having to go back to Dublin."[10]

Once Haddon's interests had turned to anthropology—in particular ethnology—the desire to record primitive races before their extinction held his attention briefly in Dublin, but he firmly believed that primitive societies were at greater risk of being lost to science than the marine organisms he lectured on at the Royal College of Science.[11] This suggests that because Haddon's anthropological thoughts revolved around *racial retrieval* rather than *racial differentiation*, they were "out of place" in Dublin's anthropometrical mission to understand Irish origins. The historiographical approach used here revealed how his ethnological interests left him in an ambivalent position regarding the role of the Anthropometric Laboratory in providing anthropological direction for Home Rule arguments in the city.

The "perception of place" and the ways in which it inflected city science, surfaced perhaps most vividly in the contrasting experiences of David Moore and Alexander Macalister. The disciplinary spaces of these two coreligionists were shown to have a considerable influence on their attitudes to Dublin. Their diverse experiences of the sites of biological investigation ultimately colored their overarching perceptions of the city, and thus figured it as either a sympathetic and enabling place or a contentious and constraining place. Take, for instance, how those Dublin spaces that could reference the "exotic and uplifting" or "domestic and familial," such as gardens, parks, stately homes, and estates, acted as a harmonious link between David Moore's convictions as a committed Presbyterian, his public role as a scientist, and his natural theological worldview. On the other hand, Alexander Macalister's network of morgues, graveyards, and cutting rooms unsettled thoughts of a beneficent designer, and his compartmentalized experiences of Dublin science—which moved between anatomical, theological, zoological, and archaeological spaces—led to a more nuanced approach to his studies on the origins and development of form.

Any examination such as this, which aims to uncover and plot "perceptions of the metropole" in a cohort of city scientists, leads to inevitable questions about the meaning of knowledge made in the city. If we accept that many "Dublins" existed in the minds of the city's scientific community, and as science has been found to be politically and culturally positioned, then new geographies of "meaning" emerge. These serve to open up the contexts of inquiry in debates on universal issues such as origins.

To illustrate this point, we need look no further than the "origins of handedness" debate that drew George Sigerson's attention in the 1880s. On closer inspection, we find that the debate was played out on two distinct ontological planes congregating around "perceptions of Dublin." Various important bodies of work relating to the origins of handedness emanated from the city during the late Victorian period. One of these, as we have seen, was Sigerson's paper to the Royal Irish Academy on dextral preeminence, in which he argued for ambidexterity in monkeys in order to downplay language links with apes. Sigerson's case for ambidexterity in apes was underpinned by a Romantic belief in a previous "Golden Age." However, a number of years after Sigerson had pronounced on the topic, Daniel Cunningham presented a paper to the Royal Anthropological Institute, "Right-Handedness and Left-Brainedness: The Huxley Lecture for 1902."[12] But for Cunningham the lateralization debate was all perfectly explicable in the context of a progressive evolution from ape to man. By pushing forward the timeline of limb evolution, he claimed, "If the ape is truly ambidextrous it is reasonable to conclude that in the evolution of man, righthandedness did not assert itself until the upper limb had been set absolutely free from the office of locomotion, and had assigned the higher duties which are now assigned to it."[13]

For the journal-reading and lecture-going scientific community, these pronouncements represented two Dublin-based academic contributions to a long-running scientific debate on handedness. However, despite Sigerson's and Cunningham's mutual interest in the topic, Cunningham made no reference to Sigerson's earlier contribution to the debate. This can, I contend, be best explained by thinking geographically about these scientists' perception of the place they inhabited and giving weight to the notion that Cunningham's "Dublin" was marked by different cognitive coordinates from Sigerson's.

Let me explain, Cunningham began his lecture by paying tribute to Thomas Huxley, who, he claimed, was "so well-known and so greatly esteemed in Ireland."[14] He thus shared with his audience a "sense of the place" where he lived and worked, which reflected a Dublin that was a hub of scientific naturalism. He then launched into a progressive evolutionary argument, which put out that "righthandedness is a character which has been attained in the ordinary course of the evolution of man by the subtle process of natural selection."[15] As was laid out in chapter 3, Cunningham's preoccupation with race fitness manifested itself as a lifelong concern with particular variations that displayed signs and symptoms of "atavism." Left-handedness for him was one such sign. So, antithetical to Sigerson's Romantic motivations, Cunningham's search for left-handedness in Dublin populations could therefore furnish his anti–Home Rule arguments with scientific evidence for degeneration and strengthen his call for strong governance from London.

These two Dublin scientists experimented, observed, and published their findings on ambidexterity in widely read medical journals. However, any claims to universality of meaning are questionable because the contention that apes were ambidextrous resulted from two very differently positioned arguments. The contexts of scientific inquiry in this debate revolved around mutually exclusive perceptions of what Dublin was and what Dublin should become. Future widening of the parameters here to include the publications of more Dublin scientists, and asking how they too conceived of their local settings and what they also took their situations to be, will allow this urban map of "scientific meaning" to be further enhanced.

Circuitries of City Science

The characters for this research were chosen across a spectrum of disciplines as good candidates for the study because of their sizable input to particular debates concerning origins. A further outcome of scrutinizing numerous lives in the city though, has been the extent to which both inter- and intradisciplinary contacts were made between the eight individuals. This serves to alert us to the myriad ways in which scientific lives also intermingle in the metropole. This is important to point out because it highlights the fact that knowledge was not confined to isolated disciplinary spaces; ideas continually crossed over from one debate to another and from one discipline to another through interacting lives. The circuitry of scientists' daily routes and commutes allowed for a diffusion of ideas throughout the city in particular ways, and offered a glimpse of how scientific ideas travel by way of conversations—within and between communities of practitioners, whether it was in an institutional setting or in less formal venues such as meeting rooms, tea shops, hostelries, city walkways, shopping streets, and parks. In the late nineteenth century, scientific endeavor was dividing into a set of discrete fields, each with practitioners, institutions, and journals to match, but the few examples from this Dublin study point to numerous incidences in which ideas cross-fertilized and diffused from place to place and from institution to institution.

The proximity and ease of connection between one scholar and another could determine how science was discussed and what kinds of interdisciplinary problems were probed. One example from the book illustrates this point. Cunningham's anatomy building was sited on a route that linked Fitzgerald's office to the rear gate of the college, and it appears that their opportunistic comings-together did indeed result in such an interdisciplinary exchange of ideas. On one occasion for instance, concepts of physiology and physics were merged in an attempt to understand the workings of the human brain. Writing to Fitzgerald in 1890, Cunningham notes, "We had a conversation some time ago about the causation of the convolutions of the brain and you put forward some mechanical views which struck me very much at the time. Within the last few weeks a very remarkable paper has appeared on the subject and I would like very much to speak to you, if you can afford me the time."[16] This is an ample illustration of the way in which *where* the scientist worked—in a city, on a campus, or in a building—mattered in terms of who was met and what was discussed. This in turn had a considerable bearing on how, and in what direction, investigations proceeded.

Diffusion of another kind occurred during these encounters also, this time not influencing merely empirical direction, but intellectual context as well—and here further notions of metropolitan cartographies come to the surface. These geographical readings do not simply point to the opportune exchange of ideas in the city that could influence the way science was done, but also reveal that entire intellectual frameworks could stand or fall depending on these social proximities.

All the personalities in this study contributed to particular branches of the origins discussion within the confines of their disciplines. But even as science was being given its disciplinary shape in this period, specialized technical endeavor in Dublin was debatably still a marginal pursuit—despite the best efforts of Fitzgerald and Barrett—in comparison with the search for wider meaning and philosophical understanding. This meant that the natural world was never far from the arena of interpretive debate, and the extent to which more profound ideas moved between disciplines and into the public domain deserves more comment. The book has brought one such topography in particular into sharper relief: the notion of a late nineteenth-century Dublin geography of "mind and matter" configurations.

The case studies have outlined the strength of the influence of Berkeleyan idealism and later Kantian transcendentalism at Trinity, and how these philosophies were worked by Trinity academics such as Fitzgerald and Macalister to both deflect and accommodate new scientific theories. The Trinity philosopher William Graham had expounded in 1872 that consciousness, not physical matter, was the first and only reality for the idealist.[17] This deep-seated dismissal of materialism in Trinity conversations can be traced back to its Divinity School, but the foregrounding of mind over matter was, by the late nineteenth century, also permeating with some vigor those disciplines with physical and intellectual proximity to the Divinity sites and savants. This is best illustrated if we circulate with our interlocutors around the environs of Trinity College, and explore some of the communicative spaces there.

The mathematical and physical science departments to which Fitzgerald was affiliated had close ties with the Divinity School. One of Fitzgerald's professional appointments at Trinity was that of registrar of the School of Engineering, which was situated in the museum building at the college.[18] The drawing school belonging to this faculty was positioned adjacent to lecture rooms used by the Divinity professors"[19] thus enabling opportunities for fortuitous encounters and exchanges of ideas. The museum building also housed the Geology School, the professor of which was the Reverend Samuel Haughton (1821–97). Haughton was a known idealist and frequently engaged with the Trinity theologians in the Divinity School. During the course of the case study discussions, it became apparent that idealist ways of thinking had commanded a strong presence in the Schools of Natural Philosophy, Geology, and Zoology and had also infiltrated the Anatomy School, where it served to buffer Macalister from encroaching materialist physiological interpretations.

Interestingly, the trajectory of these ideas can be traced in the route taken by the scientists in and out of the rear gate of Trinity College to city meeting places. For instance, on leaving the Engineering School, and taking a right-hand turn, the path veered eastward through college park, where eventually the Schools of Anatomy and Zoology were encountered at the far end of the campus before immediately exiting the university through the Lincoln Place gate. We know from Robert Lloyd Praeger's accounts that conversations "on every subject within the confines of the universe—and sometimes outside them," occurred in the tea shop near this college gate and were attended by Trinity fellows and locals alike.

Other Trinity academics ventured to Sigerson's Sunday soirees, just across the road on Clare Street.[20] Here we can see how conversations—such as that noted earlier between Fitzgerald and Cunningham—and other interdisciplinary exchanges might have influenced the direction of nascent scientific theories. Assessing the extent to which elements of Berkeleyan and Kantian philosophy and their empirical manifestations—debated over tea and toast—overflowed into nearby sites of extramural discussion—such as those of the Literary Society (6 St. Stephen's Green), the Dublin Society for Psychical Research (27 St. Stephen's Green), and the Dublin Theosophic Lodge (3 Ely Place, near St. Stephen's Green) and vice versa—would lend itself well to further geographical analysis.

A brief and coarse-grained breakdown here already provides some evidence of the "translocation and transference" of ideas across a small section of the city,[21] and issues are being raised around the ways in which Trinity philosophies—which foregrounded the mind as the "only reality"—were differently worked into new formulations as they merged with other concepts gathering on the periphery of the college. As Theodore Zeldin has pointed out, "When minds meet, they don't just exchange facts: they transform them, reshape them, draw different implications from them, [and] engage in new trains of thought."[22]

In a tribute to George Francis Fitzgerald, Oliver Lodge laments the lack of solitude for the scientist to engage in "real" work. Such men, he states, "should live apart and be preserved from the everyday struggle of common life."[23] He has to concede, however, that far from transcending the mundane, Fitzgerald's place was with "the Captains and the shouting," thus locating him well within the cut and thrust of Dublin life.[24]

In fact, we can conclude that all the scientists and scholars in this volume—Fitzgerald, Barrett, Moore, Macalister, Haddon, Cunningham, Sigerson, and MacNeill were firmly sited intellectually, socially, and culturally in the nineteenth-century city. The four spatial snapshots have illustrated how the *lived* experiences in the city, of these two physicists, a botanist, two anatomists, an ethnologist, a physician, and a historian, shaped and were in turn shaped by the spaces of science around them between 1870 and 1910. More broadly, *Geographies of City Science* has highlighted the role of Dublin city itself on the development of scientific ideas in the late nineteenth century and has shown in numerous ways how the local urban spaces of a scientist's life influenced a range of Victorian origin debates.

Notes

Introduction

Epigraph: Margulis, *Symbiotic Planet*, 89.

1. Hawking, *Brief History of Time*; Rees, *Our Cosmic Habitat*, xxi.

2. For a recent historical synopsis, see Hesketh, "Recurrence of the Evolutionary Epic."

3. J. Secord, "Introduction."

4. J. Secord, "Introduction," xlv.

5. See Shapin, "Placing the View from Nowhere"; C. Smith and Agar, "Introduction"; Livingstone, "Spaces of Knowledge"; Naylor, "Place"; Naylor, "Historical Geography"; Livingstone and Withers, "Thinking Geographically"; and Matless, "Original Theories."

6. See Livingstone, *Putting Science in Its Place*; and Finnegan, "Spatial Turn."

7. Rupke, *Alexander von Humboldt*, 9.

8. The genre of biography can be a useful tool for a contextual history of science by "opening out the individual to social and cultural contexts"; see Shortland and Yeo, *Telling Lives in Science*, 17. See also Govoni and Franceschi, *Writing about Lives.*

9. Green, "Writing Scientific Biography," 744, 745.

10. Koertge, *New Dictionary of Scientific Biography*; and Nye, "Scientific Biography," 323.

11. See Desmond and Moore, *Darwin*; and Brown, *Charles Darwin*, vols. 1 and 2.

12. Rupke, *Richard Owen, Victorian Naturalist*; see also *Richard Owen: Biology without Darwin.*

13. Westfall, *Life of Isaac Newton*; Cassidy, *Uncertainty*; Cantor, Gooding, and James, *Michael Faraday*; Holmes, *Hans Krebs*; and R. Jackson, *Ascent of John Tyndall.*

14. See Söderqvist, *History and Poetics*, 45; and Chilvers, "Tragedy of Comrade Hessen," 105, 117.

15. Söderqvist, "Existential Projects," 77.

16. Porter, "Is the Life of the Scientist a Scientific Unit?"

17. See Clifford, "Hanging Up Looking Glasses," 52; and Nye, "Scientific Biography," 322. See also J. Richards, "Introduction."

18. Terrall, "Biography as Cultural History," 308.

19. See Rupke, *Alexander von Humboldt*; Ellis, *Masculinity and Science.*

20. Shortland and Yeo, *Telling Lives in Science*, 36.

21. Green, "Writing Scientific Biography," 757.

22. Soja, *Postmetropolis*, 6.

23. Terrell, "Biography as Cultural History," 313.

24. Livingstone calls for more attention to "life geographies"; see *Putting Science in Its Place*, 182–84. Naylor too has called for more productive debates on the ways regions and places inform the theories and practices of scientists; see "Introduction," 11.

25. Rudwick, *Great Devonian Controversy.*

26. Giddens, *Modernity*, 58; for further discussion, see also Taylor, *Sources of the Self.*

27. Rorty, *Identities of Persons*, 309.

28. See Bourdieu, *Outline of a Theory.*

29. See Algazi "Scholars in Households"; Lenoir, "Scientific Habitus"; and Lahire, *l'Homme Pluriel.*

30. Lenoir, "Scientific Habitus," 42.

31. Thrift, "Remembering," 175; and Bridge, "Bourdieu."

32. Daniels and Nash, "Lifepaths," 450.

33. Mauss, "Une catégorie de l'esprit humain," 263.

34. Daston and Sibum, "Introduction," 4. For a more visceral take on this idea, see also Lawrence and Shapin's exploration of the relationship between body, mind and knowledge in *Science Incarnate.*

35. Levin et al., *Urban Modernity.*

36. Hochadel and Nieto-Galan, *Urban Histories.*

37. For recent expositions of science and Irish culture in the nineteenth century, see Adelman, *Communities of Science*; and Adelman and Agnew, *Science and Technology.* See also Bowler and Whyte, *Science and Society*; Malcolm and Jones, *Medicine, Disease and the State*; J. W. Foster, *Nature in Ireland*; and Attis and Mollan, *Science and Irish Culture.*

38. See Adelman, *Communities of Science*; Neswald, "Science and Sociability"; Leaney, "Missionaries of Science"; and J. Bennett, "Why the History of Science Matters," 9.

39. Various authors undertook syntheses of colonial science in Ireland in the 1980s and 1990s; see, for example, Jarrell, "Department of Science and Art"; Jarrell, "Colo-

nialism"; McLeod, "On Science and Colonialism"; Whyte, *Science, Colonialism and Ireland.*

40. A wider debate on "Irish" science, encompassing political and cultural factors was conducted during the 1990s in the *Irish Review.* For different readings of this topic, see Lysaght, "Themes in the History of Irish Science," 87–97; and Jones, "Catholicism."

41. See Livingstone, "Darwinism and Calvinism"; Livingstone, "Darwin in Belfast"; and Livingstone, "Science, Region and Religion." Nuala Johnson has included the Dublin Gardens in her comparative study of the historical geography of botanical gardens; see Johnson, *Nature Displaced*; and Livingstone has examined the role of Dublin-based Alexander Macalister in "Science, Text and Space," and *Dealing with Darwin.*

42. See Forgan, "But Indifferently Lodged"; and Ophir and Shapin, "Place of Knowledge," 9–12. See also Kargon, *Science in Victorian Manchester*; Inkster, *Scientific Culture*; Gooday, "Premise of Premises"; and C. Smith and Agar, "Introduction."

43. Dierig, Lachmund, and Mendelsohn, "Introduction," 18.

44. Hochadel and Nieto-Galan, *Barcelona.*

45. Begnal, *Joyce and the City*, xvi.

46. See Westwood and Williams, "Imagining Cities"; Soja, *Postmetropolis*; Frisby, *Cityscapes of Modernity*; and Rock and Moore, *Cityscapes.*

47. Soja, *Postmetropolis*, 8.

48. Torchiana, *Backgrounds for Joyce's Dubliners*, 260.

49. Gerard, *Picturesque Dublin.* O'Casey's Dublin trilogy refers to *The Shadow of a Gunman*, *Juno and the Paycock*, and *The Plough and the Stars*, the three early plays about urban life that established O'Casey's reputation. Power, *Conversations with James Joyce*, 98.

50. Whyte, "Science and Nationality," 50–52.

51. See, for instance, O'Leary, *Roman Catholicism*; Warwick, "Sturdy Protestants of Science"; and Jones, "Catholicism."

52. Livingstone and Withers, "Thinking Geographically," 7; Withers, "Place and the 'Spatial Turn.'"

53. Livingstone, *Putting Science in Its Place*, 6.

54. Rupke, "Geography of Enlightenment.

55. See Beer, "Darwin's Reading," 543; and Said, *World*, 227.

56. See Finnegan, "Placing Science," 154–55; and Livingstone, *Dealing with Darwin*, 7.

57. Livingstone and Withers, "Thinking Geographically," 7.

58. Although Catholic and Protestant neighborhoods dominated Dublin's late nineteenth-century religious landscape, a sizable Jewish community lived in an area comprising Portobello, South Circular Road, South Richmond Street, and Clanbrassil Street, collectively known as "little Jerusalem"; see Christopher, "Second City."

59. Christopher, "Second City," 151, 160.

60. For a discussion of these four analytic categories, see Lightman, *Companion to the History of Science*, 3–7.

61. Nyhart, "Historiography of Science."

62. J. Secord, "Knowledge in Transit," 660.

63. Naylor, *Regionalising Science*, 4.

64. Kohler, "Generalist's Vision."

65. Livingstone, *Putting Science in Its Place*, 7.

66. Soja, *Postmetropolis*, 6.

67. The role of the affective in science's history is yet another area that appeals to a geographical critique. Kay Anderson and Susan J. Smith have called for a fuller program of work recognizing the emotions as ways of knowing, being, and doing; see K. Anderson and Smith, "Editorial," 8.

68. Livingstone, *Putting Science in Its Place*, 180.

69. I am thinking here of Sigerson's Romantic interpretations versus Cunningham's progressive ideologies.

70. Robinson, quoted in Nelson and McCracken, *Brightest Jewel*, 150. Also Moore's sense of "a home place" has come into focus in Johnson's 2013 work on his European travels to Scandinavia and Iberia. Here, his Dublin life, or as Johnson puts it "the tastes of home" acted as a moral compass for Moore, in what was considered to be "the wider moralisation of global space" by traveling garden curators in the nineteenth century; see Johnson, "Botanical Travel."

Chapter 1: Ether

I am grateful to Routledge, Taylor and Francis for permission to reproduce some previously published material here that is drawn from my chapter in *Urban Histories of Science: Making Knowledge in the City*, ed. Oliver Hochadel and Agustí Galen-Nieto (2018).

1. Kipling, *Traffics and Discoveries*, 197–248.

2. Connor, "Transported Shiver of Bodies," 13.

3. See Sambursky, *Physical World*, 124; Hunt, "Experimenting on the Ether"; and Lodge, *Ether of Space*.

4. Einstein, "History of Field Theory."

5. Cantor and Hodge, *Conceptions of Ether*.

6. Whittaker, *History of the Theories of Aether*.

7. Schaffer, "Matter of Ether."

8. Cantor and Hodge, *Conceptions of Ether*, ix.

9. Kragh, "Sense of Crisis," 12.

10. Morus, *When Physics Became King*, 275.

11. Weaire, *George Francis Fitzgerald*, 131.

12. Mollan, "James McCullagh."

13. Preston, *Theory of Light*, 30–31.

14. For discussion of ether models, see Hunt, *Maxwellians*, 73–108.

15. Studies of the development of ether theories in nineteenth-century physics abound. A good starting place is Kragh, "Sense of Crisis," 11–14; Hunt, *Maxwellians*, 73–108; and the second half of Cantor and Hodge *Conceptions of Ether*, 170–351. For discussion on the cultural context of this development, see Morus, *When Physics Became King*, 261–87; and Brush, *Temperature of History*, 15–29.

16. Noakes, "Ethers, Religion and Politics."

17. The "Strong Program" aimed to develop a framework for the sociology of knowledge in which social factors counted not as contaminants but as constitutive of the very idea of scientific knowledge; see Shapin, "Here and Everywhere," 297.

18. Mollan, "William Fletcher Barrett"; and Mollan, "George Francis Fitzgerald."

19. W. Thomson, "On Vortex Atoms."

20. Stewart and Tait, *Unseen Universe*, 244–45.

21. For a general discussion on Thomson's vortex theory, see C. Smith and Wise, *Energy and Empire*, 419–25.

22. See Wilson, *Kelvin and Stokes*, 94.

23. Lodge, "Scope and Tendencies," 352, 354.

24. Bowler, *Reconciling Science and Religion*, 87–120.

25. McCorristine, "William Fletcher Barrett," 41.

26. Mollan, "William Fletcher Barrett," 1131.

27. *Times*, September 16, 1876, 7.

28. Cited in Noakes, "Ethers, Religion and Politics," 435. For discussion on Home Rule, see Boyce and O'Day, *Defenders of the Union*; and A. Jackson, *Home Rule*.

29. See Maguire," Socio-Economic Analysis," 40.

30. See MacVeagh, *Religious Intolerance*, 10.

31. Irish hostility to spiritualism was widespread; see Anonymous, "Psychic Night," 740.

32. See Hillier and Rooksby, *Habitus*, 11.

33. Gauld, "Barrett, Sir William Fletcher," 1.

34. Barrett, "Some Reminiscences," 282.

35. Tyndall, "Science and Spirits."

36. Lodge, *Past Years*, 120.

37. Mollan, "William Fletcher Barrett," 1114.

38. The term "sensitive flame" was coined by Tyndall and described by Barrett as "the dramatic effect of high-pitched notes on a tall and slender gas flame." The flame's behavior was comparable to that of a "sensitive nervous person uneasily starting and twitching at every little noise"; see Barrett, "Note on 'Sensitive Flames,'" 219–20.

39. *Irish Times*, January 4, 1908.

40. Ball, *Reminiscences and Letters*, 86.

41. Kelham, "Royal College of Science," 300.

42. Jones, "Scientists against Home Rule," 193–96.

43. See Livingstone, "Darwinism and Calvinism"; Livingstone, "Darwin in Belfast"; R. Foster, *Recoveries*; and O'Leary, *Roman Catholicism*, 36–37.

44. Lightman, *Origins of Agnosticism.*

45. See Jones, "Scientists against Home Rule," 193; and Luckhurst, *Invention of Telepathy*, 15.

46. Livingstone, "Darwin in Belfast."

47. Wood, "Thunderbolt of Special Wrath," 34–69.

48. Cited in Kelham, "Royal College of Science," 307; and Barrett, "Second Report of the Royal Commission," 7.

49. Ball to Barrett, October 9, 1873, cited in Noakes, "Bridge Which Is Between," 429.

50. See Kelham, "Royal College of Science," 306.

51. Anonymous, "Last Night's Conversazione."

52. Barrett, "Some Reminiscences."

53. On aims and members of societies, see Anonymous, "'Saunder's News Letter' on Spiritualism."

54. McCorristine, "William Fletcher Barrett," 42.

55. Tyndall, "Science and Spirits."

56. Barrett, "Phenomena," 937; Barrett, "Science and Spiritualism," 597.

57. Noakes, "Bridge Which Is Between," 423–29.

58. Barrett, "Light and Sound," 7. Barrett's observations on some Irish hostility to spiritualism have been recounted in a number of places; see, for instance, Anonymous, "Psychic Night"; and Anonymous, "Sir W. F. Barrett's Memories," where he mentioned that "Dublin was probably the place most antagonistic to physic matters that existed in the whole world," 1.

59. Barrett, "Sympathetic Vibrations."

60. Barrett, "Phenomena," 937.

61. Livingstone, *Putting Science in Its Place*, 182.

62. Barrett, *On the Threshold.*

63. Gauld, "Barrett, Sir William Fletcher," 2.

64. Noakes, "Ethers Religion and Politics," 431.

65. Cited in Noakes, "Ethers Religion and Politics," 435.

66. Noakes, "Ethers Religion and Politics," 435.

67. Weaire and Coey, "Mentor and Constant Friend," 515.

68. This claim, made by the *Irish Times* on July 8, 1911, at the time of the ceremonial visit of King George V and Queen Mary, reiterated the view propounded in mid-nineteenth-century guidebooks to the city, that Dublin ranked as the second city of the empire; see, for example, Fraser's *Handbook for Travellers*, 20. Christopher points out that this could be vindicated in historical terms as Dublin was the capital of England's first overseas colony; see Christopher, "Second City," 151.

69. Morus, *When Physics Became King*, 182.

70. Jones, "Scientists against Home Rule," 193.

71. See Fitzgerald, "On a Model," 417. For further discussion, see Hunt, *Maxwellians*, 73–108.

72. Fitzgerald, "Physical Science," 1–2.

73. Noakes, "Ethers, Religion and Politics," 439–44.

74. Morus, *When Physics Became King*, 269; Jones, "Scientists against Home Rule," 189.

75. For discussion, see Noakes, "Ethers, Religion and Politics," 440–41; and Warwick, "Sturdy Protestants of Science," 326–29.

76. Larmor edited the collected works of George Stokes, William Thompson, James Thomson, and G. F. Fitzgerald.

77. Purser, *As We Saw It*, 527–28; and Hunt, "Our Friend."

78. *Nature*, March 7, 1901, cited in Larmor, *Scientific Writings*, xxii.

79. Praeger, *Populous Solitude*, 192–93.

80. B. Webb, *My Apprenticeship*, 130–31.

81. Daston and Sibum, "Introduction," 3.

82. See Turner, *Contesting Cultural Authority*, 131; and D. Moore, "Deconstructing Darwinism," 375–76.

83. Moseley, "Tadpoles and Frogs."

84. C. Smith, *Science of Energy*, 170–92.

85. Luckhurst, *Invention of Telepathy*,12–17

86. Jones, "Scientists against Home Rule," 192.

87. Morus, *When Physics Became King*, 181.

88. Tyndall, *Fragments of Science*, 125.

89. Helge Kragh explains that although the supremacy of mechanics was increasingly questioned in the 1890s, there was no crisis in the mechanical worldview itself; see Kragh, "Sense of Crisis," 13.

90. Tyndall, *Address*, 45.

91. Larmor, *Aether and Matter*, 288.

92. Barrett, "Marginal Regions," 470.

93. *Record*, September 23, 1870, cited by Tyndall in *Use and Limits*, 5–6.By its critics scientific naturalism was often identified as "materialism," but apologists for the "new order" such as Tyndall and Huxley were not true materialists in the purely rationalist vein of figures such as Ray Lankester or Ludwig Buchner. Here, scientific naturalism merely implied the rigid determination of all events, including mental events, by natural laws and this meant that the only way of understanding the world was to treat it as a mechanism.

94. See Lightman, *Origins of Agnosticicm*. Barton has characterized Tyndall as a pantheist who adopted materialism as a maxim of scientific research, but not as a general philosophy; see Barton, "John Tyndall," 134.

95. Turner, *Between Science and Religion*, 20. The doctrine of "Natural Supernaturalism," propounded in 1836 by Thomas Carlyle in *Sartor Resartus*, holds that life contains elements of wonder that can never be defined or eradicated by physical science. It was an attractive concept for Victorians searching for a faith consistent with the discoveries of modern science.

96. Barton, "John Tyndall," 111.

97. Hunt, *Maxwellians*, 98–104.

98. According to the vortex atomic theory proposed by William Thomson in 1867, atoms were simply vertical structures in the continuous ether, see Thomson, "On Vortex Atoms." In this sense atoms were quasi-material rather than material bodies; see Kragh's discussion in "Sense of Crisis," 12. Stoney had conceived the electron as an ethereal singularity, rather than a charged point particle.

99. Hunt, *Maxwellians*, 100, 97.

100. Hunt, *Maxwellians*, 98.

101. Fitzgerald, "Electromagnetic Radiation," cited in Larmor, *Scientific Writings*, 276.

102. Graham, *Idealism*, 17, 20.

103. Graham, *Idealism*, 51. This view is reiterated by another Trinity scholar, Thomas Ebenezer Webb, who, in his *Veil of Isis* published in 1885, claimed that Berkeley was the first to explicitly maintain a theory of absolute idealism.

104. D'Arcy, *Adventures*, 66, 67.

105. D'Arcy, *Adventures*, 45; and Murray, *Archbishop Bernard*, 59.

106. Murray, *Archbishop Bernard*, 62.

107. For discussion of the Michelson–Morley experiment, see Goldberg and Steuwer, *Michelson Era*.

108. Fitzgerald, "Ether." 390.

109. Quoted in Hunt, *Maxwellians*, 103.

110. For the claim made at the British Association for the Advancement of Science (BAAS) meeting, see Hunt, *Maxwellians*, 102; and Fitzgerald to Heaviside August 25, 1893, cited in Hunt *Maxwellians*,103.

111. See Hunt, *Maxwellians*, 103; and Kragh, "Sense of Crisis," 13.

112. Warwick, "Sturdy Protestants of Science," 302; and Trouton, "Presidential Address," 458.

113. Livingstone, Hart, and Noll, *Evangelicals and Science*, 206.

114. Mollan, "George Frances Fitzgerald," 1215.

115. Barrett to Fitzgerald, March 5, 1898. George Francis Fitzgerald Papers, GFF 7/69.

116. Fitzgerald to Subcommittee on Fifth Book, April 19, 1893. George Francis Fitzgerald Papers, GFF 7/52.

117. Attis, "More Than a Maxwellian," 565; and for further discussion, see McDowell, *Irish Administration*, 253–56.

118. Quoted in Larmor, *Scientific Writings*, 343.

119. Fitzgerald, "Presidential Address," 558.

120. Fitzgerald to Subcommittee on Fifth Book, April 19, 1893. George Francis Fitzgerald Papers, GFF 7/52.

121. Larmor, *Scientific Writings*, 403–4.

122. Fitzgerald to Lodge, July 11, 1892. Oliver J. Lodge Papers: MS Add 89/65, cited in Noakes "Ethers Religion and Politics," 434.

123. Fitzgerald, May 7, 1896, quoted in Larmor, *Scientific Writings*, 395.

124. Attis, "More Than a Maxwellian," 569.

125. Connor, "Transported Shiver of Bodies," 10.

126. Poynting, *Collected Scientific Papers*, 264.

Chapter 2: Form

Epigraph: Russell, *Form and Function*, v.

1. Gomperz, *Greek Thinkers*, 244; Goethe's quote from "Bildung und Umbildung Organischer Naturen" is cited in Russell, *Form and Function*, 50.

2. D. W. Thompson, *On Aristotle*.

3. Nyhart, *Biology Takes Form*, 1–13.

4. Johnson, "Grand Design(er)s," 30.

5. Jones, "Darwinism in Ireland," 120.

6. Moore to Isabella Morgan, December 21, 1840. David Moore Letters, DSS/DM.

7. Ray, *Wisdom of God*, 244.

8. See Lubbock, *Prehistoric Times*; Tylor, *Primitive Culture*; and Waddell, *Foundation Myths*, 136–40.

9. For a general overview of biological thought in the late nineteenth century, see Bowler, *Evolution*; and Bowler, *Eclipse of Darwinism*. For the debate on form and function, see Amundson, "Richard Owen."

10. For responses to Darwin around the world, see Glick, *Comparative Reception of Darwinism*; Numbers and Stenhouse, *Disseminating Darwinism*; and Livingstone, *Dealing with Darwin*.

11. I am thinking here of the role of the Catholic University question in Dublin debates around Darwin; see Wood "Thunderbolt of Special Wrath," and for general discussion on Darwin and Dublin, see Jones, "Darwinism in Ireland," 117–22, and Adelman, "Evolution on Display."

12. See Ospovat, "Perfect Adaptation," 50.

13. See Desmond, *Politics of Evolution*; and Livingstone, "Natural Theology," 22.

14. Darwin, *On the Origin of Species*, 351.

15. See Ospovat's view of Huxley's evolution versus Creation categories in "Perfect Adaptation," 49–52. Ospovat prefers to categorize on the basis of biological explanation and separates naturalists into "those who advocated teleological explanations"

and those "who sought to explain organic structure and succession by biological laws and forces," 50.

16. Kohler, "Place and Practice."

17. *Irish Times*, May 29, 1860. Moore was also a member of the Royal Irish Academy (1845), and an associate (1841) and fellow (1861) of the Linnaean Society.

18. Byrne, "David Moore"; McCracken, "Botanic Gardens."

19. Byrne, "David Moore."

20. For an account of the relationship between the Botanic Gardens and the Royal Dublin Society, see Berry, *History of the Royal Dublin Society*," 186–98.

21. Moore's marriages occurred in 1836, 1847, and 1854. His first wife Hannah Bridgford of Spafield Nursery, Ballsbridge, Dublin, died of Typhus in Glasnevin in 1840, leaving him with a baby and young child. His second wife Isabella Morgan similarly died of typhus or dysentery in Glasnevin in 1847, leaving two children. He had another five children with his third wife Margaret Baker, who outlived him.

22. D. Moore, *Hand-book for the Botanic Gardens*, 16.

23. D. Moore, "On Growing."

24. Johnson, "Grand Design(er)s," 33; Paley, *Natural Theology*, 296.

25. Quoted in Desmond and Moore, *Darwin*, 449; and for discussion of Darwin's "red in tooth and claw" comment, see Lightman, *Victorian Science*, 2–3.

26. J. L. Porter, *Science and Revelation*, 13.

27. Before Tyndall's presidential address in 1874, there is little evidence to suggest that Darwin's theory was causing any profound anxiety in Calvinist Belfast; see Livingstone, "Darwin in Belfast."

28. See Johnson, "Grand Design(er)s," 29–55.

29. Smith recounts how Swinburne's "scandalous" poem "The Sundew" published in 1862 was read later in the context of Darwin's findings outlined in *Insectivorous Plants*—published the year after Moore's address—which raised anxieties about the plant's "wicked ways." According to Smith, both Darwin's book and Swinburne's poem became prominent elements in the cultural fascination that surrounded the plants from the 1860s to the 1880s. See J. Smith, "Un fleur du mal?" Moore's Belfast Address can therefore also be seen in terms of this popular debate.

30. For a detailed discussion, see J. Smith, "Grant Allen." For an overview of Darwin's evolutionary challenge to Ruskin's aesthetics, see J. Smith, *Charles Darwin*. Other discourses around Victorian representations of plants are discussed by Jackson-Houlston, "'Queen Lilies'?"

31. Haldane, "Venus Fly Trap," 107; and D. Moore, "Structure &c. of Plants," 11.

32. D. Moore, "Structure &c. of Plants," 11; and Nelson and McCracken, *Brightest Jewel*, 143.

33. Nelson and Seaward, "Charles Darwin's Correspondence."

34. D. Moore, "Structure &c of Plants," 13.

35. D. Moore, "Structure &c. of Plants." 4; and Nelson and McCracken, *Brightest Jewel*, 143.

36. Moore, "Structure &c. of Plants," 7, 9.

37. Hooker, "Address," 102–3, 116.

38. D. Moore, "Structure &c. of Plants," 19.

39. Hopkins, "Carnivorous Plants"; and D. Moore, *Hand-book for the Botanic Gardens*, 46–50.

40. D. Moore, "Structure &c. of Plants," 19. Desmond comments on the similar Paleyite views of William Buckland who speculated on the Divine skill that had gone into fashioning predators' teeth, concluding that such exquisite weapons speeded the death of the ill and infirm, increasing the aggregate health and happiness of the animals they preyed on; see Desmond, *Politics of Evolution*, 184.

41. Johnson, "Grand Design(er)s," 49.

42. See Darwin, *Various Contrivances*, and discussion by Richard Bellon, "Inspiration in the Harness."

43. D. Moore, "Structure &c. of Plants," 30.

44. D. Moore, "Structure &c. of Plants," 32.

45. For the quote from Desmond on the world of Paley, see *Politics of Evolution*, 184. Desmond uses this phrase in relation to a Paleyite Creation (*Politics of Evolution*, 184).

46. F. W. B., "In Memoriam, Charles Moore." For having been the "largest and most important contributor to the Glasnevin Botanic gardens, during the last twenty years"; see the *Irish Times*, April 3, 1863.

47. O'Brien, "Thomas Acton," 163.

48. Johnson, "Grand Design(er)s," 40–44.

49. Endersby, "Joseph Hooker," 5, 7; and Paley, *Natural Theology*, 296.

50. Latour, *Science in Action*, 215–57.

51. Livingstone, *Putting Science in Its Place*, 54. For example, Nuala Johnson has outlined the interplay of global botanical knowledge with more local concerns in Dublin, Belfast, and Cambridge, in *Nature Displaced*.

52. Ron Amundson explains that "special adaptationism"—the kind that is applied directly to species without recognition of the constraints of common descent—is an argument for special creationism but Darwin's adaptationism is not (Amundson, "Richard Owen," 35–36).

53. Lightman, *Victorian Science*, 3.

54. Darwin initially regarded his experimental botanical work as a form of relaxation and recreation; see Heslop-Harrison, "Darwin as a Botanist," 267; and Yam, Arditti, and Cameron "Orchids." Jim Endersby also discusses this in *Imperial Nature*.

55. See Endersby "Sympathetic Science."

56. Some male botanists, including Hooker, attempted to defeminize botany by professionalizing it. See Shteir, "Gender and 'Modern' Botany."

57. Hooker had penned his support for Darwin's theory as early as December 1859; see Hooker, *On the Flora of Australia*, ii.

58. Johnson, "Grand Design(er)s," 50.

59. D. Moore, *Hand-book for the Botanic Gardens*, 19, 7.

60. Stepan, *Picturing Tropical Nature*, 31–57.

61. D. Moore, *Hand-book for the Botanic Gardens*, 4.

62. D. Moore, *Hand-book for the Botanic Gardens*, 43, 51.

63. D. Moore, *Hand-book for the Botanic Gardens*, 42, 29, 60, 69.

64. D. Moore, *Hand-book for the Botanic Gardens*, 48, 22, 23, 24.

65. D. Moore, *Hand-book for the Botanic Gardens*, 51

66. D. Moore, *Hand-book for the Botanic Gardens*, 41, 34, 33, 56

67. D. Moore, *Hand-book for the Botanic Gardens*, 82; and Latour, *Science in Action*, 230.

68. D. Moore, *Hand-book for the Botanic Gardens*, 64. In Moore's outdoor departments, the herbaceous section was arranged according to De Jussieu's taxonomy; British and medicinal plants were based on De Candolle's system; and the arboretum was laid out according to the classes of Linnaeus.

69. Moore Mss, D. Moore to F. W. Moore, 6vi, 1876; and Nelson and McCracken, *Brightest Jewel*, 174. These kinds of domestic incursions into the scientific sphere are discussed by Endersby in "Sympathetic Science."

70. F. W. Moore, "Some Reminiscences"; and Nelson and McCracken, *Brightest Jewel*, 159.

71. D. Moore, *Hand-book for the Botanic Gardens*, 8.

72. Finnegan, *Natural History Societies.*

73. Ray, quoted by D. Moore in "Structure &c. of Plants," 4.

74. Quoted in Nelson and McCracken, *Brightest Jewel*, 90; and 130.

75. The policy effectively excluded the working classes from enjoying the gardens, see the *Irish Times*, January 31, 1863, and April 30, 1861.

76. Bruck and Tierney, *Landscapes of Desire*, 48.

77. *Irish Times*, January 31, 1863.

78. *Irish Times*, April 30, 1861, and April 18, 1861.

79. Hansard, House of Commons, Debate. "Opening of National Museums &c. on Sundays," June 8, 1877, cc. 1494–536. Government Hansard Reports.

80. Maurice Brooks, a member of the Irish Home Rule League and the lord mayor of Dublin in 1874, quotes Moore's opinion on Sunday opening in his correspondence, see Moore Mss. DSS/DM, June 2, 1877.

81. Robinson, "Notes on Gardens."

82. Outram, "New Spaces," 257.

83. Hansson, "Emily Lawless."

84. Lawless, "North Clare," 607; and Lysaght, *Robert Lloyd Praeger*, 13.

85. Johnson, "Grand Design(er)s," 43.

86. Johnson, "Grand Design(er)s," 29, 31.

87. Livingstone, "Science, Site and Speech," 73.

88. See J. Smith, *Charles Darwin*, 2; however, Hansson points out that a market arose for *feminized* forms of botanical texts; see "Emily Lawless," 59.

89. *Irish Times*, December 20, 1905, 8.

90. Quoted in Kelly, "Fascinating Scapel-Wielders," 509, 513.

91. Macalister, *Memoir of James Macartney*, 131–33.

92. General Medical Council, *Report of the Committee on Professional Education (1869)*, 238, quoted in Bates, "Indecent and Demoralising Representations," 21.

93. Bates, "Indecent and Demoralising Representations," 2.

94. This topic has been discussed in a series of articles in the *Dublin Historical Record* between 1987 and 1993; see, for instance, Fleetwood, "Dublin Body Snatchers: Part One"; and "Part Two"; Fleetwood, "Dublin Private Medical Schools"; and Hooper "Dublin Anatomy." For general comment on the topic, see Olry, "Body Snatchers."

95. Macalister, *Memoir of James Macartney*, 213.

96. R. Richardson, *Death, Dissection and the Destitute*, 3–30. Hurren, "Pauper Dead-House," 70.

97. Michael Sappol's work has shown that black markets operating through "official" welfare channels flourished; see Sappol, *Traffic of Dead Bodies*, 118; Hulkower also comments on this topic in "From Sacrilege to Privilege."

98. See H. MacDonald, "Procuring Corpses." The anatomists often offered incentives to those welfare agencies prepared to cooperate; see Hurren, "Pauper Dead-House"; see also Hurren and King, "'Begging for a Burial.'"

99. See R. Richardson, *Death, Dissection and the Destitute*, 193–219; Korf and Wicht, "Public Cadaver"; and Hildebrandt, "Capital Punishment."

100. Rupke, *Richard Owen: Victorian Naturalist*, 179–80.

101. Macalister, *Memoir of James Macartney*, 218.

102. See Hurren, "Pauper Dead-House," 92. For a discussion of knowledge from which lay people could be excluded on moral grounds, see Bates, "Indecent and Demoralising Representations," 2.

103. Macalister, "Review of Works on Life and Organisation," 132.

104. Jones, "Darwinism in Ireland," 117–22.

105. Barclay-Smith, "Professor Alexander Macalister," 95.

106. Fagan, "Alexander Stewart Macalister."

107. Barclay-Smith, "Professor Alexander Macalister," 97.

108. Claim made by Atheneum reviewer in 1879, quoted in Smith, "Alexander Macalister, 1844–1919," xxxvi.

109. Hayes, "Alexander Macalister."

110. Barclay-Smith, "Professor Alexander Macalister," 96.

111. Smith, "Alexander Macalister, 1844–1919," xxxix.

112. Anonymous, *Nature*, 1919.

113. Barclay-Smith "Professor Alexander Macalister," 98.

114. Smith, "Alexander Macalister, 1844–1919," xxxvi, xxxvii, xxxiv.

115. Hayes, "Alexander Macalister."

116. Macalister, "Review of *The Descent of Man*," 150. This was also contemplated by Catholic evolutionists such as St. George Mivart.

117. Macalister, "Review of *The Descent of Man*," 151.

118. Rupke, *Richard Owen: Victorian Naturalist*, 250.

119. Owen, *On the Nature of Limbs, 86.*

120. Desmond, *Archetypes and Ancestors*, 37.

121. Owen, *On the Anatomy of Vertebrates*, 807.

122. Smith, "Alexander Macalister, 1844–1919," xxxix.

123. Smith, "Alexander Macalister, 1844–1919," xxxv.

124. Lubbock's *Prehistoric Times* and Tylor's *Primitive Culture* were key texts in promoting progressive evolution in the late nineteenth century.

125. Bowler, *Evolution*, 221.

126. Macalister, "Review of *The Descent of Man*," 137.

127. Jones, "Contested Territories," 199–205.

128. Waddell, *Foundation Myths*, 106.

129. Jones, "Contested Territories," 199–205.

130. Macalister, "Review of *The Descent of Man*," 143.

131. Jones, "Contested Territories," 201.

132. Smith, "Alexander Macalister, 1844–1919," xxxvii.

133. Macalister, "Review of Works on Life and Organisation," 130.

134. Freke, *Reflections on Organisation*, 40–41. See also Blackith, *Henry Freke*, 106.

135. Macalister, "Review of Works on Life and Organisation," 131.

136. Seward and Darwin, *More Letters of Charles Darwin*, 175.

137. Rupke, *Richard Owen: Victorian Naturalist*, 205.

138. Macalister, "Review of Works on Life and Organisation," 131.

139. J. Secord, "Knowledge in Transit," 662.

140. Macalister, *Introduction to Animal Morphology*, 41.

141. Macalister, "Review of *The Descent of Man*," 140, 141, 142.

142. Macalister, "Review of Works on Life and Organisation," 129.

143. Macalister, "Bearing of Anomalous Anatomy," 197, 194.

144. Macalister, "Man Physiologically Considered," 47.

145. Fyfe, *Science and Salvation*, 117.

146. Macalister, "Man Physiologically Considered," 47.

147. Rupke, *Richard Owen: Victorian Naturalist*, 212; and Yeo, "William Whewell."

148. Rupke, "Richard Owen's Vertebrate Archetype," 249.

149. Macalister, *Evolution in Church History*, 5–38, 11.

150. Shapin, "Placing the View," 7.

151. Hurren, "Pauper Dead-House," 4.

Chapter 3: Human

I am grateful to Elsevier for permission to reproduce extracts in chapter 3 from my previously published article in *Endeavour* 39, nos. 3–4 (2015): 139–46.

1. D. J. Cunningham and Browne, "Report from the Anthropometric Laboratory," 8.

2. A four-year program of research on the Irish Ethnographic Survey is currently under way at Trinity College Dublin in association with the University of Maynooth Anthropology Department.

3. See Haddon and Browne, "Ethnography of the Aran Islands," 768–99, also later anthropometric surveys were undertaken by Browne along the west coast and published by the Royal Irish Academy between 1893 and 1900, see C. R. Brown, "Ethnography of Inisboffin and Inishark"; C. R. Brown, "Ethnography of the Mullet, Inishkea Islands, and Portnacloy"; C. R. Brown, "Ethnography of Ballycroy"; C. R. Brown, "Ethnography of Clare Island and Inishturk"; C. R. Brown, "Ethnography of Garumna and Lettermullan"; and C. R Brown, "Ethnography of Carna and Mweenish."

4. For an early synopsis of the Aran material, see Messenger, "Literary vs Scientific Interpretations." More recently Scott Ashley has discussed the blending of Romantic literary traditions with the late nineteenth–century scientific venture; see Ashley, "Poetics of Race." In 2010, Justin Carville explored the visual construction of race in "Resisting Vision."

5. For a general discussion of this topic, see Stocking, "Persistence of Polygenist Thought"; and Stocking, *Victorian Anthropology*. For a social and political critique of polygenism and monogenism, see Livingstone, *Adam's Ancestors*, 109–37. An outline of the American perspective is given by Haller in *Outcasts from Evolution*.

6. Stocking, "Persistence of Polygenist Thought," 42.

7. Jefferson, "Declaration of Independence."

8. See Price and Stewart, *To Heal the Scourge*.

9. Haller, "Species Problem," 1321; and Asma, "Metaphors of Race," 17.

10. Jowett, *Politics of Aristotle*, 11.

11. Kuklick and Kohler, "Introduction," 9.

12. Dain, *Hideous Monster*, 247.

13. Livingstone, *Adam's Ancestors*, 172.

14. See Levy and Peart, "Secret History," 2.

15. See Bigelow, *Fiction, Famine*; see also R. Richards, "Darwin's Place," 10059; and Hoppen, *Mid-Victorian Generation*, 223.

16. Kingsley, *Charles Kingsley*, 242–24; and Hoppen, *Mid-Victorian Generation*, 224.

17. Greg, "Realities of Irish Life," 78.

18. For further discussion on this, see Bowler, "Race Theory and the Irish," 144; and Kuklick, *Savage Within*, 118.

19. See Renan, *La poésie des races celtiques.*

20. Nicholas, "Why Are the Methodists of Ireland?" 351, 352.

21. Kohler, "Lab History," 786.

22. Rich, *Race and Empire*, 104.

23. J. P. Jackson and Weidman, *Race, Racism and Science*, 72.

24. See Johann Friedrich Blumenbach, *Anthropological Treatises*, 16.

25. Blumenbach and Petrus Camper, in particular. See Haller, *Outcasts from Evolution*, 5–7. For a discussion of Petrus Camper's anthropological legacy, see Meijer, *Race and Aesthetics.*

26. Gould, *Mismeasure of Man*, 86; and C. R. Brown, "Report of the Work."

27. Gould, *Mismeasure of Man*, 74.

28. Latour, *Pandora's Hope*, 58.

29. Trinity College Muniment Records, TCD Mun. V5 vol. 2.

30. Dublin University Calendar 1891. Papers belonging to D. J. Cunningham. Trinity College Medical School Records Mun/Med/V3.

31. Flower, *Essays on Museums*, 260.

32. D. J. Cunningham and Haddon, "Anthropometric Laboratory of Ireland," 36–37. See also Forrest's outline in the "Anthropometric Laboratory of Ireland."

33. Haddon Papers, 3058.

34. *The Book of Trinity College Dublin* (1892).

35. See Forrest, "Anthropometric Laboratory of Ireland," for further discussion of the application of Galton's methodologies in Dublin, 1384.

36. Galton, "On Recent Designs," 2.

37. Willey and Sabloff, cited in Finger and Fernando, "E. George Squier," 361, 362.

38. Rothenberg, "Joseph Henry Papers."

39. Stocking, "Persistence of Polygenist Thought," 45.

40. Haller, *Outcasts from Evolution*, 23.

41. Haddon and Brown, "Ethnography of the Aran Islands," 783–91.

42. Gould, *Mismeasure of Man*, 86.

43. See Gould, *Mismeasure of Man*, 5; and Morton, *Crania Americana.*

44. Gould, *Mismeasure of Man*, 51

45. Flower, *Essays on Museums*, 260; Haddon Papers, 4018.

46. Trinity College Muniment Records, Mun V5, vol 2, 1891–99.

47. See Galton, "On Recent Designs," 5; Grey and Robinson, "Auricular Head Height Techniques," 324; and C. R. Browne, "Some New Anthropological Instruments."

48. C. R. Browne, "Some New Anthropological Instruments."

49. Bowler, *Evolution*, 283; and Keith, *Antiquity of Man*, 151.

50. Meijer, *Race and Aesthetics*; White, *Account of the Regular Gradation in Man*, 67.

51. Haller, *Outcasts from Evolution*, 19–40.

52. Haller, *Outcasts from Evolution*, 29.

53. See A. Thompson, "Influence of Posture"; and Shute, "Racial Anatomical Peculiarities," 124.

54. Haller, *Outcasts From Evolution* 49.

55. D. J. Cunningham, *Lumbar Curve in Man.*

56. A. Thompson, "Daniel John Cunningham," 98.

57. Cullen, "Pure School of Science."

58. Barrett, *Historical Sketch*, 7.

59. Jones, "Darwinism in Ireland," 119.

60. See McCorristine, "Science and Nation," 3–4; and Sigerson, *Modern Ireland*, 201.

61. D. J. Cunningham and Haddon, "Anthropometric Laboratory of Ireland," 36. One of Haddon's primary guiding texts when he first embarked on the Anthropometric Survey was John Beddoe's controversial *The Races of Britain.* Here, the Celt and Anglo-Saxon were differently placed by an "Index of Nigrescence."

62. Stocking, *Race, Culture and Evolution*, 57; Topinard, *Man in Nature.*

63. Hartley, "1890 Correspondence Book," quoted in McCorristine, "Science and Nation," 7

64. Haddon, *History of Anthropology*, 47.

65. Rich, *Race and Empire*, 104.

66. Macalister, *Some Morphological Lessons*; and Gillham, "Evolution by Jumps."

67. Haddon, "Revision of the British Actiniae."

68. See Jones, *Social Darwinism*, 146, and see Urry's essay, "From Zoology to Ethnology," esp. 66–68.

69. These ideas are outlined in Haddon, "Six Lectures on Anthropology."

70. Haddon, *Introduction to the Study of Embryology*, 40.

71. Haller, *Outcasts from Evolution*, 40.

72. Haddon, "Reports on the State of Science," 33; and Haddon, "Saving of Vanishing Knowledge."

73. Haddon, *Races of Man.*

74. Haddon and Browne, "Ethnography of the Aran Islands," 781.

75. Scott Ashley and Greta Jones have covered this topic, see Ashley, "Poetics of Race"; and Jones, "Contested Territories."

76. Rich, *Race and Empire*, 105–6.

77. Haddon, "Presidential Address."

78. See D. J. Cunningham, *Lumbar Curve in Man*; D. J. Cunningham "Spinal Curvature"; D. J. Cunningham, "Occasional Eighth True Rib; D. J. Cunningham, "Significance of Anatomical Variations"; D. J. Cunningham, "Varying Forms of the Stomach"; and D. J. Cunningham, "Head of an Aboriginal Australian."

79. See D. J. Cunningham, "Alleged Physical Deterioration"; D. J. Cunningham, "Anniversary Address"; and D. J. Cunningham, "Address to Prime Minister."

80. Galton, "Possible Improvement of the Human Breed"; Papers and Correspondence of Daniel John Cunningham, D4.

81. D. J. Cunningham, "Scotch Radical's Experience," 164.

82. For Cunningham's views on the physical mean, see "Alleged Physical Deterioration," 156; and D. J. Cunningham, "Scotch Radical's Experience," 164.

83. D. J. Cunningham, "Scotch Radical's Experience ," 166.

84. Turpin, "John Hogan," 64.

85. *Irish Ecclesiastical Record*, March 1884.

86. See L. Spencer, "Chance, Circumstance and Folly," 8.

87. D. J. Cunningham, "Address to Prime Minister.," 425.

88. D. J. Cunningham, "Significance of Anatomical Variations."

89. D. J. Cunningham, "Significance of Anatomical Variations," 2.

90. D. J. Cunningham, "Significance of Anatomical Variations."

91. D. J. Cunningham, "Skeleton of the Irish Giant"; D. J. Cunningham and Bennett, "Brain and Eyeball"; D. J. Cunningham, "Skull and Some of the Other Bones"; D. J. Cunningham, *On the Brain of the Microcephalic Idiot*; and D. J. Cunningham, "On The Microcephalic Brain."

92. D. J. Cunningham, "Spinal Curvature," 489.

93. Scobie, " Long Road Home."

94. D. J. Cunningham, "Spinal Curvature," 489.

95. D. J. Cunningham, "On the Sacral Index."

96. Haller, "Species Problem," 1321.

97. See Urry, "Notes and Queries"; and Urry, "Englishmen, Celts and Iberians."

98. It was only through the persistence of their assistant C. R. Browne, that the survey results were published when the lab closed in 1898; see Forrest, "Anthropmetric Laboratory of Ireland," 1384.

99. D. J. Cunningham, "Anthropometric Investigations," 156–58; and Haddon, "Manners and Customs," 638.

100. Hinks, McKivigan, and Williams, *Encyclopedia of Antislavery*, 327.

Chapter 4: Word

I am grateful to Elsevier for permission to reproduce extracts in chapter 4 from my previously published article in *Endeavour* 39, nos. 3–4: 2015: 139–46.

1. Duddy, "Thinking Ireland," 15.

2. For explorations of the more general relationship between science and nationalism, see Crawford *Nationalism and Internationalism in Science*, where the argument is made that late nineteenth-century science was an integral part of the emergence of the nation-state in Europe; also Duddy, *History of Irish Thought*.

3. Bauman and Briggs, *Voices of Modernity*.

4. Robins, *Short History of Linguistics*, 130; and Harvey, *Condition of Postmodernity*, 9.

5. Harland, *Literary Theory*, 87; and Hobsbawn, *Age of Extremes*, 179.

6. von Humboldt, *"On Language,"* 46.

7. See Duranti, *Companion to Linguistic Anthropology*, 374; and for a lengthier discussion, see Baumann and Briggs, *Voices of Modernity*, 139–204.

8. Mathews, *Revival*, 8.

9. See B. Anderson, *Imagined Communities*; and Johnson, "Making Space."

10. Mathews, *Revival*, 9.

11. See Mac Giolla Chriost, *Irish Language in Ireland*.

12. Hyde, "Necessity for De-Anglicising Ireland."

13. Bucholtz and Hall, "Language and Identity," 374.

14. Hirsch, "Imaginary Irish Peasant," 1120. Patten comments on this further in "Ireland's 'Two Cultures Debate.'"

15. Allan, "States of Mind," 151.

16. For a broader look at nineteenth-century links between Irish and Indian nationalism, see Foley and O'Connor, *Ireland and India*.

17. See Eco, *Search for the Perfect Language*, 17.

18. In regard to this, StephenAlter draws attention to a contemporary predilection for conceptual transfer between disciplines, particularly between biology and linguistics; see Alter, *Darwinism*, 108–46.

19. See P. Richardson, "Consolation of Philology," 1–13.

20. Bopp, *Comparative Grammar*, 1251, 1443.

21. Radick, *Simian Tongue*; and Radick, "Race and Language."

22. Alter, *Darwinism*, 10.

23. For further discussion, see Bauman and Briggs, *Voices of Modernity*, 163–97; De Beaugrande, "Closing the Gap," 424; and Uitti, *Linguistics and Literary Theory*.

24. House of Commons debate, February 18, 1884, vol. 284, cc. 1179–80. Government Hansard Reports.

25. Ken McGilloway's biography of Sigerson is the most comprehensive account to date; see McGilloway, *George Sigerson*. Other biographical sketches can be found in Fahie, "Dr. George Sigerson"; and Lyons, "Medicine and Literature"; Lyons also produced an article for the *Journal of the History of the Neurosciences* (see "George Sigerson"), and compiled the entry on Sigerson for McGuire and Quinn, *Dictionary of Irish Biography*.

26. House of Commons debate, February 18, 1884, vol. 284, cc. 1179–80; see Lyons, "George Sigerson," 50.

27. See McGeachie, "Science, Politics," 130–33.

28. McGeachie, "Science, Politics," 121; and Geary, *Medicine and Charity.*

29. Lyons, "George Sigerson," 51; Sigerson's death notice in the *Irish Times* in 1925 also mentions that his "Fenian friendships and associations must for a long time have left him somewhat solitary among the Dublin doctors"; see McGilloway, *George Sigerson*, 100; and Anonymous, "George Sigerson: An Obituary."

30. See A. Cunningham and Jardine, "Introduction," in *Romanticism and the Sciences*; Knight, "Romanticism and the Sciences"; R. Richards, *Romantic Conception of Life*; and Tresch, *Romantic Machine*. For a glimpse of Sigerson's Romantic persona, see Anonymous, "Irish Sword."

31. R. Richards, *Romantic Conception of Life*, 5.

32. See Curran, "Oisin after the Fianna."; Doolin, *Dublin's Medical Schools*, 73; and McGilloway, *George Sigerson*, 25–31.

33. Fahie, "Dr. George Sigerson," 59.

34. There was a considerable overlap between membership of the Literary Society and membership of both the Dublin Theosophic Lodge and the Society for Psychical Research (SPR) in Dublin; see McCorristine, "William Fletcher Barrett," 48. The Dublin branch of the SPR had its inaugural meeting in the nearby Shelbourne Hotel, at No. 27 St. Stephen's Green.

35. Anonymous, "New Members."

36. Early examples include Douglas Hyde's folklore collection *Beside the Fire* in 1890, and his musical compilation *The Love Songs of Connaught* in 1893. Yeats's gathering of folklore, myth, and legend in *The Celtic Twilight* was also published in 1893.

37. Tierney, "What Did the Gaelic League Accomplish?" 344; MacNeill, "Some Notes of Our National Literature"; Eoin McNeill Papers, unpublished memoirs. LAI/Q (vi), 496.

38. Naylor terms this the "charisma of place," in "Place," 9.

39. Anonymous, "Prefatory."

40. Sigerson, "Genesis and Evolution," 20.

41. Sigerson, "Irish Literature," 64, 69.

42. Sigerson, "Genesis and Evolution," 20, 24.

43. Müller, *Three Lectures*, 6.

44. For a discussion of this aspect of the revival, see Collins, *Catholic Churchmen*; for Sigerson"s medical circles, see also Meenan, "Catholic University School": for comment on Archbishop Walsh's role in promoting the school, 141, and the Nationalist outlook of the students and staff, 140–43.

45. Bourke, *Aryan Origin*, 299, 109.

46. Sigerson, "Irish Literature," 75.

47. Sigerson, "Irish Literature," 72, 73.

48. Sigerson, "Genesis and Evolution," 24–25.

49. See Kropotkin, *Mutual Aid.*

50. In *Bards of Gael and Gall,* Sigerson describes the coming together of two civilizations in mutual literary appreciation—Irish and Elizabethan; see Deane, "George Sigerson," 238; and Sigerson, *Last Independent Parliament.*

51. Mivart, *Man and Apes,* 188.

52. Hunter describes these events in *Makings of Dr. Charcot's Hysteria Shows,* 12.

53. Ellenberger, *Discovery of the Unconscious,* 95.

54. Goldstein sees it as part of a wider European political project to challenge the influence of religion; see Goldstein, "Hysteria Diagnosis."

55. Binet and Féré, *Animal Magnetism,* 86.

56. Robertson, "Hypnotism at Paris and Nancy," 494.

57. Harrington, "Metals and Magnets."

58. Marshall, "Archaeology of the Abstract Body," 95.

59. See Goldstein, "Hysteria Diagnosis."

60. Harrington, "Metals and Magnets," 23.

61. See Sigerson, "Micro-Atmospheric Researches."

62. Harrington, "Metals and Magnets."

63. Sigerson, " Examination," 183.

64. See Sigerson, "Examination"; and Sigerson, "Influence of Solonoids."

65. Marshall, "Archaeology of the Abstract Body," 98.

66. Charcot, "Lecture on Certain Phenomena," 790.

67. Yeats, *Celtic Twilight,* 61, 63.

68. A. Cunningham and Jardine, "Introduction," in *Romanticism and the Sciences* 2–3.

69. See Pick, *Faces of Degeneration.*

70. See Hyde, "Necessity for De-Anglicising Ireland," 118.

71. Gilman and Chamberlain, *Dark Side of Progress.*

72. Quoted in Lyons, "George Sigerson," 53.

73. Anonymous, "Reviews and Bibliographical Notices."

74. Marshall has explored Charcot's use of the archetype in "Archaeology of the Abstract Body."

75. A. H. Bennett, "Metalloscopy and Metallotherapy." However, the Cecilia Street School was more amenable to French influence than other Dublin medical institutions. F. O. C. Meenan recounts that the school Golden Jubilee dinner was attended by a visiting French naval squadron, and the evening ended with a rendition of *La Marseillaise*; see Meenan, "Catholic University School."

76. Finger, *Origins of Neurosciences,* 37–38, 375–78; and Broca, "Remarques sur le siège de la faculté"; see also Corballis, "Gestural Origins of Language."

77. Radick, *Simian Tongue.*

78. See Tylor, *Researches into the Early History*, 54; and Lubbock, *Prehistoric Times*, 32.

79. Sigerson, "On Alternate Paralysis"; Sigerson, "Examination"; and Sigerson, "Consideration."

80. Harris, "On the Evolution of Handedness," 148–49.

81. Broca, "Remarques sur le siège de la faculté."

82. Ferrier, "Experimental Researches."

83. Ogle, "On Dextral Pre-eminence," 292.

84. Pye-Smith, "Left-handedness."

85. Darwin, *Variation of Animals*, 399.

86. See Sigerson, "Consideration," 40, 50, 4.

87. Sigerson, "Consideration," 37. This was a method Sigerson learned from Duchenne de Boulogne (49).

88. Sigerson, "Consideration," 40.

89. Bateman, *On Aphasia*, 107–8.

90. Sigerson, "Consideration," 41.

91. Sigerson, "Influence of Solonoids."

92. Sigerson, "Consideration," 43. Sigerson was alerted to this phenomenon by Rev. Dr. Samuel Haughton.

93. Ogle, "On Dextral Pre-eminence," 287.

94. See Deniker, "Les gorillas vivant."

95. Sigerson, "Consideration," 43n17.

96. Aarsleff, *From Locke to Saussure*, 170–71.

97. Sigerson, "Irish Literature," 68.

98. Ryan, "Eoin MacNeill," 433.

99. In later life, he accused historians who equated the progress of civilization with the growth of state power of "worshiping the Beast and his image"; see Maume and Edwards, "Eoin MacNeill."

100. Later published in Duffy, Sigerson, and Hyde, *Revival of Irish Literature.*

101. See MacCartney, "MacNeill and Irish Ireland," 86; Le Roux, *L'Irlande militante.*

102. The authenticity of Macpherson's Gaelic translations were questioned by other scholars, including Samuel Johnson; see Mulholland, "James Macpherson's Ossian Poems."

103. MacNeill, *Phases of Irish History*, 8; and MacNeill, "Vindication."

104. Gibbons, "Constructing the Canon," 954.

105. MacNeill, "Irish Nation"; and MacNeill, "Race and Nationality."

106. Balfour, *Nationality and Home Rule*, 13, 15.

107. O'Tuama, *Gaelic League Idea*, 98.

108. Kuper, *Ancestors.*

109. MacNeill, *Early Irish Laws*, 89.

110. Laird, "India."

111. O'Kelly, *Guide*, 3; and MacNeill, *Early Irish Laws*, 91–94, 84–86.

112. Binchy, "MacNeill's Study," 43–48, see also MacNeill, "Irish Law of Dynastic Succession." Quoted in Binchy, "MacNeill's Study," 45.

113. MacNeill, "Ni h-Iad Na Mna Breaghtha Ghnios," 18.

114. Bowler, *Evolution*, 304.

115. MacNeill, "Our Whig Inheritance," 64; see also MacNeill, "Irish Nation."

116. Mathews, *Revival*, 23. For synopses on the work of the league, see also Cleary, "Gaelic League"; and Tierney, "What Did the Gaelic League Accomplish?"

117. See West, *Horace Plunkett*, 20–36.

118. See McMahon's survey, "All Creeds and All Classes?" in *Grand Opportunity*, 85–127.

119. O'Casey, *Drums under the Window*, 10.

120. McMahon, *Grand Opportunity*, 109, 111.

121. MacNeill, "Ni h-Iad Na Mna Breaghtha Ghnios," 17, 18.

122. *An Claidheamh Soluis*, March 18, 1899.

123. *Gaelic League Pamphlet*, May 12–15, 1903, 9–10; see Moran, "Battle of Two Civilisations."

124. Plunkett, *Ireland in the New Century*, 148–49.

125. Moran, "Battle of Two Civilisations," 110–11.

126. Mathews, *Revival*, 28.

127. Williams, "Irish Astronomical Tract," 90.

128. MacNeill, "Ni h-Iad Na Mna Breaghtha Ghnios," 18, 19.

129. See Daston and Stolleis, *Natural Law*, 1–13.

130. MacNeill's legal training was evident in a number of his publications on early Irish history; see, for instance, *Early Irish Laws*, and "Ireland and Wales."

131. Eoin McNeill Papers, LAI/Q (iii), 573.

132. Whewell, *History of the Inductive Sciences*, 483–84.

133. Eoin McNeill Papers, LAI/Q (iii), 573. See also MacNeill's book, *Phases of Irish History*, in which he uses this methodological approach in discussing the different epochs of Irish history.

134. Alter, *Darwinism*, 11.

135. French, "Linguistic Science," 612.

136. Irvine and Gal, "Language Ideology."

137. Amsterdamska, "Institutions and Schools of Thought."

138. MacNeill was instrumental in introducing new phonetic sound laws to Irish-language studies; see Hamp, "MacNeill–O'Brien Law."

139. Quoted in Robins, *Short History of Linguistics*, 183.

140. MacNeill, "Modern Standard of Irish," 181.

141. See Robins, *Short History of Linguistics*, 183; MacNeill, "Study of Irish," 188; and MacNeill, "Modern Standard of Irish," 181.

142. MacNeill, "Ni h-Iad na Mna Breaghtha Ghnios," 17.

143. MacNeill, "Study of Irish," 171–72.

144. MacNeill, "Some Points in Spelling," 262–63. He discussed these ideas further in a number of other short articles; see, for instance, "Irish Phonetics"; "Method in Teaching Irish" 9; and "On the Irish Infinitive."

145. French, "Linguistic Science," 612.

146. MacNeill, "Some Points in Spelling," 262–63.

147. Eoin McNeill Papers, 1868–1979, LAI/C/3, 4, quoted in French, "Linguistic Science," 614.

148. MacNeill, "Some Points in Spelling," 263.

149. MacNeill, "Ni h-Iad Na Mna Breaghtha Ghnios," 18; and MacNeill, "Some Points in Spelling," 263.

Epilogue: Locating Origins in Dublin

1. Abbott, *Flatland.*

2. See Randall, *Warped Passages.*

3. Livingstone, *Putting Science in Its Place*, 6.

4. See Dennis, *Cities in Modernity*, 69.

5. Shapin, "Placing the View," 4–7.

6. Livingstone and Withers, *Geographies of Nineteenth-Century Science*, 13.

7. Livingstone, *Putting Science in Its Place*, 180.

8. Haddon, *Reports of the Cambridge Anthropological Expedition*, 1:xii.

9. Haddon to Frazer, October 25, 1899, cited in Urry "From Zoology to Ethnology," 80.

10. Fanny Haddon to Alfred Cort Haddon,1898. Alfred Cort Hadden Papers, 12.

11. Haddon asked Fitzgerald to deal with the loose ends of his Committee work in Dublin, because of a pressing need to return to Cambridge to prepare for imminent "tropical trips." Haddon to Fitzgerald, January 27, 1888. George Francis Fitzgerald Papers, GFF 23/57.

12. D. J. Cunningham, "Right-Handedness."

13. D. J. Cunningham, "Right-Handedness," 286.

14. D. J. Cunningham, "Right-Handedness," 273.

15. D. J. Cunningham, "Right-Handedness," 284, and he also noted, "It thus comes about that it is in civilised races engaged in skilled labour of the highest order that the highest degree of righthandedness is exhibited." He moved on to argue, on what he admitted to be shaky evidence, that there was a great reduction in the percent-

age of those who were right-handed among "microcephalic idiots"; see Cunningham, "Right-Handedness," 286.

16. Cunningham to Fitzgerald, May 21, 1890. George Francis Fitzgerald Papers GFF 19/2.

17. Graham, *Idealism.*

18. The museum building was completed in 1857.

19. Bourke, "Early Buildings."

20. Fahie, "Dr. George Sigerson," 59.

21. This concept is explored by Livingstone in his discussion on the ways in which scientific knowledge moves from place to place; see Livingstone, *Putting Science in Its Place*, 140–47.

22. Zeldin, *Conversation*, 14.

23. Lodge, quoted in Larmor, *Scientfic Writings*, xix.

24. Lodge, quoted in Larmor, *Scientfic Writings*, xxi.

Bibliography

Archive Collections

Cambridge University Library, Manuscripts Department: Alfred Cort Haddon Papers: 12; 22; 23; 3058; 4018; 4046; 4062; 5048; 5090; 5350; 5351.

Centre For Research Collections Edinburgh University: Papers and Correspondence of Daniel John Cunningham: GB 0237; Gen. 2003; A3/C2-C7; C5; C6; C7; D3; D4; D5.

Manuscripts and Archives Research Library, Trinity College, Dublin: Charles R. Browne Photograph Collection: TCD MS 10961; Trinity College Muniment Records TCD Mun /V/3; TCD/Mun /V/5; Trinity College Medical School Records Mun/ Med/V3.

National Botanic Gardens Library, Glasnevin, Dublin: David Moore Letters, DSS/ DM; GLASRA Periodicals and National Botanic Gardens Occasional Papers.

National Library of Ireland, Dublin: George Sigerson Correspondence and Notes: MS 8100; MS G103; MS 17,771; Eoin MacNeill Correspondence: MS 10,883.

Royal Dublin Society Library, Dublin: George Francis Fitzgerald Papers: GFF 2/1–90; 7/9–134; 15/1–18; 19/2; 23/57.

Royal Irish Academy Library, Dublin : Royal Irish Academy Muniments: SR/18/1A-2B.

UCD Archives, Department of History and Archives, University College, Dublin: Eoin McNeill Papers: UCDA LAI/ A-R; LA30/PH/325; Royal College of Science Papers: UCDA: RCSI/212-248; Correspondence Book UCDA: RCSI/28

Government Hansard Reports.

Opening of National Museums, &c on Sundays.—Resolution. House of Commons Debate, June 8, 1877, vol. 234, cc. 1495–1536.

Prisons (Ireland) Commission—Dr. Sigerson. House of Commons Debate, February 18, 1884, vol. 284, cc. 1179–80.

Contemporary Newspapers and Periodicals

American Anthropologist
American Journal of Physical Anthropology
An Claidheamh Solais
Brain
British Journal of Psychiatry
British Medical Journal
Bulletins et Mémoires de la Société Anatomique de Paris
Contemporary Review
Dublin Historical Record
Dublin Journal of Medical Science
Dublin Quarterly Journal of Medical Science
Dublin Review
Freeman's Journal
Gaelic Journal
Gardeners' Chronicle
Good Words
Ireland American Review
Ireland To-Day
Irish Ecclesiastical Record
Irish Press Agency
Irish Statesman
Irish Sword
Irish Theological Quarterly
Irish Times
Irish Unionist Alliance
Journal of Anatomy
Journal of Anatomy and Physiology
Journal of the Kew Guild
Journal of the National Literary Society
Journal of the Royal Anthropological Institute of Great Britain and Ireland
La Nature: Revue des Sciences et de Leur Applications aux Arts et a L'industrie,
Leader
Light
Man
Nation
National Student
Nature
New Ireland Review
New York Times
Nineteenth Century
Nonconformist
Philosophical Magazine

Plain Words
Popular Science
Present Day Tracts
Proceedings of the Royal Irish Academy
Proceedings of the Royal Society of Edinburgh
Proceedings of the Royal Society of London
Proceedings of the Society for Psychical Research
Quarterly Journal of Science
Science
Science Progress
Scientific Proceedings of the Royal Dublin Society
Scientific Transactions of the Royal Dublin Society
Times
Transactions of the Medical and Chirurgical Society of London
Transactions of the Royal Academy of Medicine in Ireland
Transactions of the Royal Irish Academy
Transactions of the Royal Society of Edinburgh
Western Argus

Primary and Secondary Sources

Aarsleff, H. *From Locke to Saussure: Essays on the Study of Language and Intellectual History.* London: Athlone, 1982.

Abbott, E. *Flatland: A Romance of Many Dimensions.* London: Seeley & Co., 1884.

Adelman, J. *Communities of Science in Nineteenth-Century Ireland.* Science and Culture in the Nineteenth Century 10. London: Pickering and Chatto, 2009.

Adelman, J. "Evolution on Display: Promoting Irish Natural History and Darwinism at the Dublin Science and Art Museum." *British Journal for the History of Science* 38, no. 4 (2005): 411–36.

Adelman, J., and A. Agnew, eds. *Science and Technology in Nineteenth-Century Ireland.* Dublin: Four Courts Press, 2011.

Algazi, G. "Scholars in Households: Refiguring the Learned Habitus 1480–1550." *Science in Context* 16, no. 1/2 (2003): 9–42.

Allan, N. "States of Mind: Science, Culture and the Irish Intellectual Revival 1900–1930." *Irish University Review Special Issue: New Perspectives on the Irish Literary Revival* 31, no. 1 (Spring–Summer 2003): 150–64.

Alter, S. *Darwinism and the Linguistic Image: Language, Race and Natural Theology in the Nineteenth Century.* Baltimore: Johns Hopkins University Press, 1999.

Amsterdamska, O. "Institutions and Schools of Thought: The Neogrammarians." *American Journal of Sociology* 91, no. 2 (September 1985): 332–58.

Amundson, R., ed. "Richard Owen and Animal Form," in R. Owen, *On the Nature of Limbs: A Discourse, with Introductory Essays by Amundson, R., Padian, K., Winsor, M., and Coggon, J.* Chicago: University of Chicago Press, 2007.

Anderson, B. *Imagined Communities: Reflections on the Origins and Spread of Nationalism.* London: Verso, 1991.
Anderson, K., and S. J. Smith. "Editorial: Emotional Geographies." *Transactions of the Institute of British Geographers* 26, no. 1, New Series (2001): 7–10.
Anonymous. "George Sigerson M.D., F.R.C.P.I., M.R.I.A.: An Obituary." *British Medical Journal,* March 7, 1925.
Anonymous. "An Irish Sword for Marshall MacMahon." *Irish Sword* 4, no. 17 (1980): 246–48.
Anonymous. "Last Night's Conversazione." *Freeman's Journal,* August 16, 1878: 5.
Anonymous. *Nature* (1919): 26.
Anonymous. "New Members." *Journal of the National Literary Society* 1 (1900): xv–xxii.
Anonymous. "Prefatory." *Journal of the National Literary Society* 1 (1900): i.
Anonymous. "A Psychic Night at the Authors' Club: Sir William Barrett on Psychical Research." *Light* 43 (1923): 740–41.
Anonymous. "Reviews and Bibliographical Notices." *Dublin Journal of Medical Science* 74 (1882): 201–2.
Anonymous. "'Saunder's News Letter' on Spiritualism." *Proceedings of the Society for Psychical Research* (1882): 39–41.
Anonymous. "Sir W. F. Barrett's Memories." *Western Argus* 24, no. 5046 (January 22, 1924): 1–36.
Ashley, S. "The Poetics of Race in 1890s Ireland: An Ethnography of the Aran Islands." *Patterns of Prejudice* 35, no. 2 (2001): 5–18.
Asma, S. "Metaphors of Race: Theoretical Presuppositions behind Racism." *American Philosophical Quarterly* 32, no. 1 (1995): 13–29.
Attis, D. "More Than a Maxwellian: Fitzgerald and Technology." *European Review* 15, no. 4 (2007): 561–73.
Attis, D., and C. Mollan, eds. *Science and Irish Culture: Why The History of Science Matters in Ireland.* Dublin: The Royal Dublin Society, 2004.
B. F. W. "In Memoriam, Charles Moore." *Journal of the Kew Guild* 13, no. 2 (1905): 264–66.
Balfour, A. J. *Nationality and Home Rule.* London: Longmans Green and Co., 1913.
Ball, R. *Reminiscences and Letters of Sir Robert Ball.* Cassell and Co., 1915.
Barclay-Smith, E. "Professor Alexander Macalister 1844–1919." *Journal of Anatomy* 54 (October 1919): 96–99.
Barrett, W. F. *An Historical Sketch of the Royal College of Science From Its Foundation to the Year 1900.* Dublin, 1907.
Barrett, W. F. "Light and Sound: An Examination of Their Reputed Analogy." *Quarterly Journal of Science* 1 (1870): 1–16.
Barrett, W. F. "The Marginal Regions of Science." *Contemporary Review* 24 (1913): 467–76.
Barrett, W. F. "Note on 'Sensitive Flames.'" *Philosophical Magazine* 33 (1867)3: 219–20.
Barrett, W. F. *On the Threshold of the Unseen: An Examination of the Phenomena of Spiritualism and of the Evidence of Life after Death.* London: Kegan Paul, 1917.

Barrett, W. F. "The Phenomena of Spiritualism." *Nonconformist* 36 (1875): 934–37.

Barrett, W. F. "Science and Spiritualism." *Light* 13 (1894): 539–97.

Barrett, W. F. Cited in "Second Report of the Royal Commission on Technical Instruction: vol. 4, evidence &c., relating to Ireland." *Parliamentary Papers* 21 (1884): 75.

Barrett, W. F. "Some Reminiscences of Fifty Years in Psychical Research." *Proceedings of the Society for Psychical Research* 34 (1924): 275–95.

Barrett, W. F. "Sympathetic Vibrations." *Good Words* 32 (1891): 46.

Barton, R. "John Tyndall, Pantheist: A Rereading of the Belfast Address." *Osiris* 2, no. 3 (1987): 111–34.

Bateman, F. *On Aphasia, or Loss of Speech in Cerebral Disease.* London: J. E. Adlard, 1868.

Bates, A. W. "Indecent and Demoralising Representations: Public Anatomy Museums in Mid-Victorian England." *Medical History* 52 (2008): 1–22.

Bauman, R., and C. L. Briggs. "Language Philosophy as Language Ideology: John Locke and Johann Gottfried Herder." In *Regimes of Language: Ideologies, Polities, and Identities*, ed. P. V. Kroskity, 139–204. Santa Fe: School of American Research Press, 2000.

Bauman, R., and C. L. Briggs. *Voices of Modernity: Language Ideologies and the Politics of Inequality.* Cambridge: Cambridge University Press, 2003.

Beddoe, J. *The Races of Britain: A Contribution to the Anthropology of Western Europe.* London: Trübner and Co., 1885.

Beer, G. "Darwin's Reading and the Fictions of Development." In *The Darwinian Heritage*, ed. D. Kohn, 543–88. Princeton, NJ: Princeton University Press, 1985.

Begnal, M., ed. *Joyce and the City: The Significance of Place.* New York: Syracuse University Press, 2002.

Bellon, R. "Inspiration in the Harness of Daily Labor: Darwin, Botany and the Triumph of Evolution." *Isis* 102, no. 3 (2011): 393–420.

Bennett, A. H. "Metalloscopy and Metallotherapy." *Brain* 1, no. 3 (1878): 331–39.

Bennett, J. "Why the History of Science Matters in Ireland." In Attis and Mollan, *Science and Irish Culture*, 1–15.

Berry, H. F. *A History of the Royal Dublin Society.* London: Longmans Green, 1915.

Bigelow, G. *Fiction, Famine and the Rise of Economics in Victorian Britain and Ireland.* Cambridge: Cambridge University Press, 2004.

Binchy, D. A. "MacNeill's Study of the Ancient Irish Laws." In Martin and Byrne, *Scholar Revolutionary*, 37–48.

Binet, A. and C. Féré. *Animal Magnetism.* New York: D. Appleton, 1888.

Blackith, R. E. "Henry Freke: An Early Dublin Evolutionist." *Irish Naturalists' Journal* 20, no. 3 (July 1980): 104–6.

Blumenbach, J. F. *The Anthropological Treatises of Johann Friedrich Blumenbach.* London: Anthropological Society, 1865.

The Book of Trinity College Dublin. 1591–1891. Belfast: Marcus Ward, 1892.

Bopp, F. *A Comparative Grammar of the Sandscrit, Zend, Greek, Latin, Lithuanian, Gothic, German and Slavonic Languages.* London: James Madden, 1853.

Bourdieu, P. *An Outline of a Theory of Practice.* Cambridge: Cambridge University Press, 1977.

Bourke, U. T. *The Aryan Origin of the Gaelic Race and Language.* London: Longmans Green, 1875.

Bourke, U. T. "The Early Buildings." *The Book of Trinity College Dublin 1591–1891,* 183–235. Dublin: Hodges, Figgis, 1892.

Bowler, P. *The Eclipse of Darwinism: Anti-Darwinian Evolution Theories in the Decades around 1900.* Baltimore: Johns Hopkins University Press, 1983.

Bowler, P. *Evolution: The History of an Idea.* 3rd ed. Berkeley: University of California Press, 2003.

Bowler, P. *Life's Splendid Drama.* Chicago: University of Chicago Press, 1996.

Bowler, P. "Race Theory and the Irish." In *Social Thought on Ireland in the Nineteenth Century,* ed. S. O'Siochain, 135–66. Dublin: University College Dublin Press, 2009.

Bowler, P. *Reconciling Science and Religion: The Debate in Early Twentieth Century Britain.* Chicago: University of Chicago Press, 2001.

Bowler, P., and N. Whyte, eds. *Science and Society in Ireland: The Social Context of Science and Technology in Ireland 1800–1950.* Dublin: Institute of Irish Studies, 1997.

Boyce, D. G., and A. O' Day, eds. *Defenders of the Union: A Survey of British and Irish Unionism since 1800.* London: Routledge, 2001.

Bridge, G. "Bourdieu, Rational Action and the Time–Space Strategy of Gentrification." *Transactions of the Institute of British Geographers* 26, New Series (2001): 205–24.

Broca, P. "Remarques sur le siége de la faculté du langage articulé"; Suivies d'une observation d'aphemie (Perte de la Parole). *Bulletins et Mémoires de la Société Anatomique de Paris* 36 (1861b): 330–57.

Brooke, J. *Science and Religion: Some Historical Perspectives.* Cambridge: Cambridge University Press, 1991.

Brown, J. *Charles Darwin. Voyaging.* Vol. 1. Princeton, NJ: Princeton University Press, 1996.

Brown, J. *Charles Darwin. The Power of Place.* Vol. 2. New York: Alfred A. Knopf, 2002.

Browne, C. R. "The Ethnography of Ballycroy, Co. Mayo." *Proceedings of the Royal Irish Academy* 4 (1896): 74–111.

Browne, C. R. "The Ethnography of Carna and Mweenish, in the Parish of Moyruss, Connemara." *Proceedings of the Royal Irish Academy* 6 (1900): 503–34.

Browne, C. R. "The Ethnography of Clare Island and Inishturk, Co. Mayo." *Proceedings of the Royal Irish Academy* 5 (1897): 40–72.

Browne, C. R. "The Ethnography of Garumna and Lettermullan, in the County of Galway." *Proceedings of the Royal Irish Academy* 5 (1898): 223–68.

Browne, C. R. "The Ethnography of Inisboffin and Inishark, Co. Galway." *Proceedings of the Royal Irish Academy* 3 (1893): 317–70.

Browne, C. R. "The Ethnography of the Mullet, Inishkea Islands, and Portnacloy, Co. Mayo." *Proceedings of the Royal Irish Academy* 3 (1895): 587–649.

Browne, C. R. "Report of the Work Done in the Anthropometric Laboratory of Trinity College, from 1891 to 1898." *Proceedings of the Royal Irish Academy* 5 (1898): 269–93.

Browne, C. R. "Some New Anthropological Instruments." *Proceedings of the Royal Irish Academy* 2, third series (1892): 397–99.

Bruck J., and A. Tierney. *Landscapes of Desire: Parks, Colonialism and Identity in Victorian and Edwardian Ireland.* Dublin: Heritage Council, 2009.

Brush, S. *The Temperature of History, Phases of Science and Culture in the Nineteenth Century.* New York: Burt Franklin, 1978.

Bucholtz, M., and K. Hall. "Language and Identity." In *A Companion to Linguistic Anthropology*, ed. A. Duranti, 369–94. London: Wiley Blackwell, 2004.

Buchner, L. *Force and Matter: Empirico-Philosophical Studies, Intelligibly Rendered.* London: Trübner, 1864.

Byrne, P. "David Moore." *Dictionary of Irish Biography*, accessed October 15, 2013, http://dib.cambridge.org.queens.ezp1.qub.ac.uk/quicksearch.do.

Cantor, G. N., D. Gooding, and F. James. *Michael Faraday.* London: Humanity Books, 1996.

Cantor, G. N., and M. J. S. Hodge. *Conceptions of Ether: Studies in the History of Ether Theories, 1740 to 1900.* Cambridge: Cambridge University Press, 1981.

Carville, J. "Resisting Vision: Photography, Anthropology and the Production of Race in Ireland." In *Visual, Material and Print Culture in Nineteenth-Century Ireland*, ed. C. Breathnach and C. Lawless, 158–75. Dublin: Four Courts Press, 2010.

Cassidy, D. C. *Uncertainty, the Life and Science of Werner Heisenberg.* London: Palgrave Macmillan, 1994.

Charcot, J. M. "A Lecture on Certain Phenomena of Hysteria Major, Delivered at the Salpetriere, 17 November, translated by George Sigerson." *British Medical Journal* 11 (1878): 789–91.

Chilvers, C. A. J. "The Tragedy of Comrade Hessen: Biography as Historical Discourse." In *The History and Poetics of Scientific Biography*, ed. T. Söderqvist, 105–21. Science, Technology and Culture 1700–1945 Series. Aldershot: Ashgate, 2007.

Christopher, A. J. "The Second City of the Empire: Colonial Dublin 1911." *Journal of Historical Geography* 23, no. 2 (1997): 151–63.

Cleary, A. E. "The Gaelic League 1893–1919." *Studies: An Irish Quarterly Review* 8, no. 31 (September 1919): 398–408.

Clifford, J. "Hanging Up Looking Glasses at Odd Corners: Ethnobiographical Prospects." In *Studies in Biography*, ed. D. Aaron, 41–57. Harvard English Studies 8. Cambridge, MA: Harvard University Press.

Collins, K. *Catholic Churchmen and the Celtic Revival in Ireland, 1848–1916.* Dublin: Four Courts Press, 2002.

Connor, S. "Transported Shiver of Bodies: Weighing the Victorian Ether." *British Association of Victorian Studies*, 2004, accessed March 1, 2014, http://www.stevenconnor.com/ether/.

Corballis, M. "The Gestural Origins of Language." *American Scientist* 87, no. 2 (March–April 1999): 138–41.

Crawford, E. *Nationalism and Internationalism in Science 1880–1939.* Cambridge: Cambridge University Press, 1992.

Cullen, C. "'A Pure School of Science': The Royal College of Science for Ireland and Scientific Education in Victorian Ireland." Paper presented at Science and Technology in Nineteenth-Century Ireland. Society for Nineteenth-Century Ireland, Annual Conference, 2009.

Cunningham, A., and N. Jardine, eds. *Romanticism and the Sciences.* Cambridge: Cambridge University Press, 1990.

Cunningham, D. J. "Address to Prime Minister Henry Campbell Bannerman as Part of the Deputation of Associated Societies on a Proposed National Anthropometric Survey." *Journal of the Royal Anthropological Institute of Great Britain and Ireland* 37 (July–December 1907): 424–32.

Cunningham, D. J. "The Alleged Physical Deterioration of the People." Report to Anthropology at the British Association Cambridge meeting on Anthropometric Investigation in Great Britain and Ireland, August 17th to 24th 1904." *Man* 4 (1904): 155–60.

Cunningham, D. J. "Anniversary Address." *Journal of the Royal Anthropological Institute of Great Britain and Ireland* 38 (January–June 1908): 10–35.

Cunningham, D. J. "Anthropometric Investigations in Britain and Ireland." *Report of the British Association for the Advancement of Science*, Seventy-Fourth Meeting, Cambridge, 1904.

Cunningham, D. J. "The Head of an Aboriginal Australian." *Journal of the Royal Anthropological Institute of Great Britain and Ireland* 37 (January–June 1907): 47–57.

Cunningham, D. J. *The Lumbar Curve in Man and the Apes, with an Account of the Topographical Anatomy of the Chimpanzee, the Orang-Utan and Gibbon.* Dublin: Royal Irish Academy (Cunningham Memoirs 2), 1886.

Cunningham, D. J. "The Occasional Eighth True Rib in Man and Its Relation to Right Handedness." *Journal of Anatomy and Physiology* 24, no. 1 (1889): 127–29.

Cunningham, D. J. *On the Brain of the Microcephalic Idiot.* Dublin: Royal Dublin Society, Scientific Transactions, vol. 5, series 2, 1895.

Cunningham, D. J. "On the Microcephalic Brain: Communicated by D. J. Cunningham M.D. F.R.S. to the Anthropological Section of the BAAS, Bradford September 1900." *Journal of the Anthropological Institute of Great Britain and Ireland* 30 (1900): 104.

Cunningham, D. J. "On the Sacral Index: Communicated by D. J. Cunningham MD FRS to the Anthropological Section of the BAAS, Bradford September 1900." *Journal of the Anthropological Institute of Great Britain and Ireland* 30 (1900): 97.

Cunningham, D. J. "Report of the Committee of Investigation in the Anatomical Department of Trinity College Session 1888–1889." *Transactions of the Royal Academy of Medicine in Ireland* 7, no. 1 (December 1889): 378–87.

Cunningham, D. J. "Right-Handedness and Left-Brainedness: The Huxley Lecture for 1902." *Journal of the Anthropological Institute of Great Britain and Ireland* 32 (1902): 273–96.

Cunningham, D. J. "A Scotch Radical's Experience of Ireland." *Irish Unionist Alliance* 2 (1903): 163–67.
Cunningham, D. J. "The Significance of Anatomical Variations." *Journal of Anatomy and Physiology* 33, no. 13 (1898): 2–9.
Cunningham, D. J. "The Skeleton of the Irish Giant Cornelius McGrath." *Transactions of the Royal Irish Academy* 29 (1891): 553–612.
Cunningham, D. J. "The Skull and Some of the Other Bones of the Skeleton of Cornelius MaGrath, the Irish Giant." *Journal of the Royal Anthropological Institute of Great Britain and Ireland* 21 (1892): 40–41.
Cunningham, D. J. "The Spinal Curvature in an Aboriginal Australian." *Proceedings of the Royal Society of London* 45 (1889): 487–504.
Cunningham, D. J. "The Varying Forms of the Stomach in Man and the Anthropoid Ape." *Transactions of the Royal Society of Edinburgh* 45, no. 1 (1906): 9–47.
Cunningham, D. J., and E. H. Bennett. "The Brain and Eyeball of a Cyclopian Monster." *Transactions of the Royal Irish Academy* 29 (1891): 101–22.
Cunningham, D. J., and C. R. Browne. "Report from the Anthropometric Laboratory of Trinity College, Dublin, on the Head Measurements of the Anatomists who attended the Meeting of the Anatomical Society in Dublin on the 10th and 11th of June 1898." *Journal of Anatomy and Physiology* 33 (1898/1899): 4–8.
Cunningham, D. J., and A. C. Haddon. "The Anthropometric Laboratory of Ireland." *Journal of the Anthropological Institute of Great Britain and Ireland* 21 (1892): 35–39.
Curran, C. P. "Oisin after the Fianna." *Irish Statesman*, February 28, 1925.
D'Arcy, C. F. *Adventures of a Bishop*. London: Hodder and Staughton, 1934.
Dain, B. *A Hideous Monster of the Mind: American Race Theory in the Early Republic*. Cambridge, MA: Harvard University Press, 2002.
Daniels, S., and C. Nash. "Lifepaths: Geography and Biography." *Journal of Historical Geography* 30 (2004): 449–58.
Darwin, C. R. *The Descent of Man and Selection in Relation to Sex*. London: John Murray, 1871.
Darwin, C. R. *Insectivorous Plants*, London: John Murray, 1875.
Darwin, C. R. *On the Origin of Species*. Edited and with an Introduction and Notes by Gillian Beer. Oxford: Oxford University Press, 1996.
Darwin, C. R. *On the Origin of Species by Means of Natural Selection, or the Preservation of Favoured Races in the Struggle for Life*. London: John Murray, 1859.
Darwin, C. R. *The Variation of Animals and Plants under Domestication*. London: John Murray, 1868.
Darwin, C. R. *The Various Contrivances by which Orchids Are Fertilized by Insects, and on the Good Effects of Intercrossing*. London: John Murray, 1862.
Daston, L., and H. O. Sibum. "Introduction: Scientific Personae and Their Histories." *Science in Context* 16, no. 1/2 (2003): 1–8.
Daston, L., and M. Stolleis, eds. *Natural Law and the Laws of Nature in Early Modern Europe: Jurisprudence, Theology, Moral and Natural Philosophy*. Aldershot: Ashgate, 2008.

De Beaugrande, R. "Closing the Gap between Linguistics and Literary Study: Discourse Analysis and Literary Theory." *Journal of Advanced Composition* 13, no. 2 (1993): 423–48.

Deane, S. "Eoin Mac Neill." In Deane *Field Day Anthology*, 2:285–88.

Deane, S., ed. *The Field Day Anthology of Irish Writing*. 3 vols. Derry: Field Day, 1991.

Deane, S. "George Sigerson." In Deane, *Field Day Anthology*, 2:238–48.

Deniker, J. "Les gorillas vivant à Paris." *La Nature: Revue Des Sciences et de Leur Applications aux Arts et a L'industrie* 19 (1891): 369–70.

Dennis, R. *Cities in Modernity: Representations and Productions of Metropolitan Space 1840–1930*. Cambridge: Cambridge University Press, 2008.

Desmond, A. *Archetypes and Ancestors: Palaeontology in Victorian London 1850–1875*. London: Blond and Briggs, 1982.

Desmond, A. *The Politics of Evolution: Morphology, Medicine and Reform in Radical London*. Chicago: University of Chicago Press. 1992.

Desmond, A., and J. Moore. *Darwin*. London: Penguin, 1992.

Dierig, S., J. Lachmund, and J. A. Mendelsohn. "Introduction: Toward an Urban History of Science." *Osiris* 18 Second Series (2003): 1–19.

Doolin, W. *Dublin's Medical Schools: A Biographical Retrospect*. Dublin: Welcome Foundation, 1952.

Duddy, T. *A History of Irish Thought*. London: Routledge, 2002.

Duddy, T. "Thinking Ireland: Cultural Nationalism and the Problem of Irish Ideas." *New Hibernia Review* 7, no. 1 (Spring 2003): 14–23.

Duffy, C. G., G. Sigerson, and D. Hyde, eds. *The Revival of Irish Literature and Other Addresses*. London: T. Fisher, 1894.

Duranti, A., ed. *A Companion to Linguistic Anthropology*. Oxford: Blackwell, 2004.

Easton, H. *A Treatise on the Intellectual Character and Civil and Political Condition of the Coloured People of the United States*. Boston: Issac Knapp, 1837.

Eco, U. *The Search for the Perfect Language*. Oxford: Blackwell, 1995.

Einstein, A. "The History of Field Theory ('Olds and News of Field Theory')." *New York Times*, February 3, 1929.

Ellenberger, H. *The Discovery of the Unconscious, the History and Evolution of Dynamic Psychiatry*. London: Allen Lane, 1970.

Ellis, H. *Masculinity and Science in Britain 1831–1918*. London: Palgrave Macmillan, 2017.

Endersby, J. *Imperial Nature: Joseph Hooker and the Practices of Victorian Science*. Chicago: University of Chicago Press, 2008.

Endersby, J. "Joseph Hooker, the Making of a Botanist." *Endeavour* 25, no. 1 (2001): 1–7.

Endersby, J. "Sympathetic Science: Charles Darwin, Joseph Hooker and the Passions of Victorian Naturalists." Special issue: "Darwin and the Evolution of Victorian Studies." *Victorian Studies* 51, no. 2 (Winter 2009): 299–320.

Fagan, B. "Alexander Stewart Macalister (1870–1950)." *Oxford Dictionary of National Biography*, accessed September 11, 2013, http://www.oxforddnb.com.queens.ezp1.qub.ac.uk/view/article/57475.

Fahie, N. "Dr. George Sigerson." *Dublin Historical Record* 38 (1985): 53–60.

Ferrier, D. "Experimental Researches in Cerebral Physiology and Pathology." *Journal of Anatomy and Physiology* 8, no. 1 (1873): 152–55.

Finger, S. *The Origins of Neuroscience: A History of Explorations into Brain Function*. Oxford: Oxford University Press, 2001.

Finger, S., and H. R. Fernando. "E. George Squier and the Discovery of Cranial Trepanation: A Land Mark in the History of Surgery and Ancient Medicine." *Journal of the History of Medicine* 56 (October 2001): 353–81.

Finnegan, D. A. *Natural History Societies and Civic Culture in Victorian Scotland*. London: Routledge, 2009.

Finnegan, D. A. "Natural History Societies in Late Victorian Scotland and the Pursuit of Local Civic Science." *British Journal for the History of Science* 38, no. 1 (2005): 53–72.

Finnegan, D. A. "Placing Science in an Age of Oratory: Spaces of Scientific Speech in Mid-Victorian Edinburgh." In Livingstone and Withers, *Geographies of Nineteenth-Century Science*, 153–77.

Finnegan, D. A. "The Spatial Turn: Geographical Approaches to the History of Science." *Journal of the History of Biology* 41, no. 2 (2008): 369–88.

Finnegan, D. A., and J. J. Wright. *Spaces of Global Knowledge: Exhibition, Encounter and Exchange in an Age of Empire*. Oxford: Routledge, Taylor and Francis, 2015.

Fitzgerald, G. F. "Electromagnetic Radiation." *Proceedings of the Royal Institution* 13 (1890): 77–85.

Fitzgerald, G. F. "The Ether and the Earth's Atmosphere." *Science* 13 (1889): 390.

Fitzgerald, G. F. "On a Model Illustrating Some Properties of the Ether." *Scientific Proceedings of the Royal Dublin Society* 4 (1885): 407–19.

Fitzgerald, G. F. "Physical Science and Its Connections." *Science Progress* 1 (1894): 1–11.

Fitzgerald, G. F. "Presidential Address to Section A—Mathematical Sciences." *Report of the Fifty-Eighth Meeting of the British Association for the Advancement of Science, Bath 1888*, 557–62. London: John Murray, 1889.

Fleetwood, J. F. "The Dublin Body Snatchers: Part One." *Dublin Historical Record* 42, no. 1 (December 1988): 32–40.

Fleetwood, J. F. "The Dublin Body Snatchers: Part Two." *Dublin Historical Record* 42, no. 2 (March 1989): 42–52.

Fleetwood, J. F. "Dublin Private Medical Schools in the Nineteenth Century." *Dublin Historical Record* 46, no. 1 (Spring 1993): 31–45.

Flower, W. H. *Essays on Museums and other Subjects Connected with Natural History*. London: Macmillan, 1898.

Foley, T., and M. O'Connor. *Ireland and India: Colonies, Culture and Empire*. Dublin: Irish Academic Press, 2006.

Forgan, S. "But Indifferently Lodged . . . Perception and Place in Building for Science in Victorian London." In Smith and Agar, *Making Space for Science*, 195–215.

Forrest, D. "The Anthropometric Laboratory of Ireland." *American Psychologist* 41, no. 12 (1986): 1384–85.

Foster, J. W., ed. *Nature in Ireland: A Scientific and Cultural History.* Dublin: Lilliput Press.

Foster, R. *Recoveries: Neglected Episodes in Irish Cultural History.* Dublin: University College Dublin Press, 2002.

Fraser, J. *A Handbook for Travellers in Ireland.* Dublin: Curry's Handbooks, 1844.

Freke, H. *Reflections on Organisation: or, Suggestions for the Construction of an Organic Atomic Theory.* Dublin: J. McGlashan, 1848.

French, B. "Linguistic Science and Nationalist Revolution: Expert Knowledge and the Making of Sameness in Pre-Independence Ireland." *Language in Society* 38 (2009): 607–25.

Frisby, D. *Cityscapes of Modernity: Critical Explorations.* London: Polity, 2001.

Fyfe, A. *Science and Salvation: Evangelical Popular Science Publishing in Victorian Britain.* Chicago: University of Chicago Press, 2004.

Galton, F. "The Possible Improvement of the Human Breed under the Existing Conditions of Law and Sentiment." *Man* 1 (1901): 161–64.

Galton, F. "On Recent Designs for Anthropometric Instruments." *Journal of the Anthropological Institute* 16 (1887): 2–8.

Garrigan Mattar, S. *Primitivism, Science and the Irish Revival.* Oxford: Clarendon Press, 2004.

Gauld, A. "Barrett, Sir William Fletcher (1844–1925)." *Oxford Dictionary of National Biography.* Oxford: Oxford University Press, 2004.

Geary, L. *Medicine and Charity in Ireland 1718–1851.* Dublin: University College Dublin Press, 2005.

Gerard, F. A. *Picturesque Dublin: Old and New.* London: Hutchinson, 1898.

Gibbons, L. "Constructing the Canon: Versions of National Identity." In Deane, *Field Day Anthology,* 2:950–55.

Giddens, A. *Modernity and Self-Identity: Self and Society in the Late Modern Age.* Cambridge: Polity Press, 1991.

Gillham, N. W. "Evolution by Jumps: Francis Galton, William Bateson and the Mechanism of Evolutionary Change." In "Perspectives: Anecdotal, Historical and Critical Commentaries on Genetics," ed. J. F. Crow and W. F. Dove. *Genetics* 159 (December 2001): 1383–92.

Gilman, S., and E. Chamberlain. *The Dark Side of Progress.* New York: Columbia University Press, 1985.

Glick, T. F., ed. *The Comparative Reception of Darwinism.* Austin: University of Texas Press, 1972.

Goldberg, S., and R. Steuwer, eds. *The Michelson Era in American Science 1870–1930.* New York: American Institute of Physics, 1988.

Goldstein, J. "The Hysteria Diagnosis and the Politics of Anti-Clericalism in Late Nineteenth-Century France." *Journal of Modern History* 54, no. 2 (1982): 209–39.

Gomperz, T. *Greek Thinkers: A History of Ancient Philosophy.* Vol. 4. Translated by G. Berry. London: John Murray, 1905.

Gooday, G. "Nature in the Laboratory: Domestication and Discipline with the Microscope in Victorian Life Sciences." *British Journal for the History of Science* 24 (1991): 307–41.

Gooday, G. “Placing or Replacing the Laboratory in the History of Science?” Focus: Laboratory History, *Isis* 99, no. 4 (2008): 783–95.

Gooday, G. “The Premise of Premises: Spatial Issues in the Historical Construction of Laboratory Credibility.” In Smith and Agar, *Making Space for Science*, 216–45.

Gould, S. *The Mismeasure of Man*. London: Penguin, 1981.

Gould, S. *Ontogeny and Phylogeny*. Cambridge, MA: Belknap Press of Harvard University Press, 1977.

Govoni, P., and Z. A. Franceschi. *Writing about Lives in Science: (Auto) Biography, Gender and Genre*. Göttingen: V&R Unipress, 2014.

Graham, W. *Idealism: An Essay Metaphysical and Critical*. London: Longmans, Green, 1872.

Green, M. T. “Writing Scientific Biography.” *Journal of the History of Biology* 40, no. 4 (2007): 727–59.

Greg, W. R. “Realities of Irish Life.” *Quarterly Review* 126 (1869): 61–80.

Grey, H., and S. O. Robinson. “Auricular Head Height Techniques with a Comparison of OH by Todd’s Head Spanner and Hrdlicka’s Compass Held up, on the Living.” *American Journal of Physical Anthropology* 11 (1928): 301–10.

Haddon, A. C. *A History of Anthropology*. London: Watts, 1910.

Haddon, A. C. *An Introduction to the Study of Embryology*. London: Blackiston, 1887.

Haddon, A.C. “Manners and Customs of the Torres Straits Islanders.” *Nature* 42 (1890): 637–42.

Haddon, A. C. “Presidential Address: Anthropology Its Position and Needs.” *Journal of the Anthropological Institute* 33 (1903): 11–13.

Haddon A. C. *The Races of Man and Their Distribution*. Cambridge: Cambridge University Press, 1929.

Haddon, A. C. *Reports of the Cambridge Anthropological Expedition to the Torres Straits*. 6 vols. Cambridge: Cambridge University Press, 1901.

Haddon, A.C. “Reports on the State of Science. Section H.” *Report of the Sixty-Seventh Meeting of the British Association for the Advancement of Science, held at Toronto*, 1897.

Haddon, A. C. “A Revision of the British Actiniae Part 1.” *Science Transactions of the Royal Dublin Society* 2, no. 5, second series (1889): 297–361.

Haddon, A. C. “The Saving of Vanishing Knowledge.” *Nature* 55, no. 1422 (1897): 305–6.

Haddon, A. C. “Six Lectures on Anthropology given at the Royal College of Science, Dublin.” Science and Art Department Pamphlet. February–March 1893, Alfred Cort Haddon Papers 3058.

Haddon, A. C., and C. R. Browne. “The Ethnography of the Aran Islands, Co. Galway.” *Proceedings of the Royal Irish Academy* 2, New Series, 1893.

Haldane, J. B. S. “Venus Fly Trap.” *Nature* 5 (June 1874): 105–7.

Haller J. S. *Outcasts from Evolution: Scientific Attitudes of Racial Inferiority, 1859—1900*. Carbondale: Southern Illinois University Press, 1971.

Haller, J. S. “The Species Problem: Nineteenth-Century Concepts of Racial Inferiority in the Origin of Man Controversy.” *American Anthropologist* 72, no. 6, New Series (1970): 1319–29.

Hamp, E. P. "The MacNeill–O'Brien Law." *Eriu* 25 (1974): 172–80.

Hansson, H. "Emily Lawless and Botany as a Foreign Science." *Journal of Literature and Science* 4, no. 1 (2011): 59–73.

Harland, R. *Literary Theory from Plato to Barthes: An Introductory History.* New York: St. Martin's Press, 1996.

Harrington, A. "Metals and Magnets in Medicine: Hysteria, Hypnosis and Medical Culture in Fin-de-Siècle Paris." *Journal of Psychological Medicine* 18, no. 1 (1988): 21–38.

Harris, L. J. "On the Evolution of Handedness: A Speculative Analysis of Darwin's Views and a Review of Early Studies of Handedness in the 'Near Allies of Man.'" *Brain and Language* 73 (2000): 132–88.

Hartley, W. "1890 Correspondence Book." *UCD Archives* RCSI/28.

Harvey, D. *The Condition of Postmodernity.* Oxford: Blackwell, 1989.

Hawking, S. *A Brief History of Time: From Big Bang to Black Holes.* New York: Bantam, 1989.

Hayes, C. "Alexander Macalister." *Dictionary of Irish Biography*, accessed October 7, 2013, http://dib.cambridge.org.queens.ezp1.qub.ac.uk/viewReadPage.do?articleId=a5092&searchClicked=clicked&quickadvsearch=yes.

Hesketh, I. "The Recurrence of the Evolutionary Epic." *Journal of the Philosophy of History* 9, no. 2 (2015): 196–219.

Heslop-Harrison, J. "Darwin as a Botanist." in *A Century of Darwin*, ed. S. A. Barnett, 267–96. London: Mercury Books, 1962.

Hildebrandt, S. "Capital Punishment and Anatomy: History and Ethics of an On-Going Association." *Clinical Anatomy* 21, no. 1 (January 2008): 5–14.

Hillier, J., and E. Rooksby, eds. *Habitus: A Sense of Place.* Aldershot: Ashgate, 2005.

Hinks, P., J. R. McKivigan, and R. Williams. *Encyclopedia of Antislavery and Abolition.* Westport, CT: Greenwood Press, 2007.

Hirsch, E. "The Imaginary Irish Peasant." *Proceedings of the Modern Language Association* 106, no. 5 (1991): 1116–33.

Hobsbawn, E. *The Age of Extremes: The Short Twentieth Century, 1914–1991.* London: Abacus, 1994.

Hochadel, O., and A. Nieto-Galan. *Barcelona: An Urban History of Science and Modernity 1888–1929.* Oxford: Routledge Taylor and Francis, 2016.

Hochadel, O., and A. Nieto-Galan. *Urban Histories of Science: Making Knowledge in the City 1820–1940.* Oxford: Routledge Taylor and Francis, 2019.

Holmes, F. L. *Hans Krebs, Volume 1: The Formation of a Scientific Life 1900–1933.* Oxford: Oxford University Press, 1993.

Hooker, J. "Address to the Department of Botany and Zoology." In *Report of the Forty-Fourth Meeting of the British Association for the Advancement of Science held at Belfast in August 1874*, 102–16. London: John Murray, 1875.

Hooker, J. *On the Flora of Australia: Its Origins, Affinities and Distribution. Being an Introductory Essay to the Flora of Tasmania*, parts 1–3. London: Lovell Reeve, 1859.

Hooper, A. C. "Dublin Anatomy in the 17th and 18th Centuries." *Dublin Historical Record* 40, no. 4 (September 1987): 122–32.

Hopkins, E. "Carnivorous Plants." *Contemporary Review* 35 (1879): 50.
Hoppen, K. T. *The Mid-Victorian Generation 1846–1886.* Oxford: Oxford University Press, 1998.
Hulkower, R. "From Sacrilege to Privilege: The Tale of Body Procurement for Anatomical Dissection in the United States." *Einstein Journal of Biology and Medicine* 27, no. 1 (2011): 23–26.
Hunt, B. J. "Experimenting on the Ether: Oliver J. Lodge and the Great Whirling Machine." *Historical Studies in the Physical and Biological Sciences* 16, no. 1 (1986): 111–34.
Hunt, B. J. *The Maxwellians.* Ithaca, NY: Cornell University Press, 1991.
Hunt, B. J. "Our Friend of Brilliant Ideas: G. F. Fitzgerald and the Maxwellian Circle." Focus: Fitzgerald and the Maxwellians, *European Review* 15, no. 4 (2007): 531–44.
Hunter, D. *The Makings of Dr. Charcot's Hysteria Shows.* New York: Edwin Mellen Press, 1998.
Hurren, E. "A Pauper Dead House: The Expansion of the Cambridge Anatomical Teaching School under the late Victorian Poor Law, 1870–1914." *Medical History* 48, no. 1 (January 2004): 69–94.
Hurren, E., and S. King. "'Begging for a Burial': Form, Function and Conflict in Nineteenth-Century Pauper Burial." *Social History* 30, no. 3 (August 2005): 321–41.
Hutchinson, J. *The Dynamics of Cultural Nationalism: The Gaelic Revival and the Creation of the Irish Nation State.* London: Allan and Unwin, 1987.
Huxley, T. H. *Lay Sermons, Addresses and Reviews.* London: Macmillan,1871.
Huxley, T. H. "Owen's Position in the History of Anatomical Science." In Rev. Owen, *Life of Professor Owen.* Vol. 2. London, 1894, 273–332.
Hyde, D. *Beside The Fire: A Collection of Irish Gaelic Folk Stories.* London: David Nutt, 1890.
Hyde, D. *The Love Songs of Connaught.* Dundrum: Dun Emer Press, 1904.
Hyde, D. "The Necessity for De-Anglicising Ireland." In Duffy, Sigerson, and Hyde, *Revival of Irish Literature*, 115–54.
Inkster, I. *Scientific Culture and Urbanisation in Industrialising Britain.* Varioram Collected Study Series. Vol. 602. Farnham: Ashgate, 1997.
Irvine, J., and S. Gall. "Language Ideology and Linguistic Differentiation." In *Regimes of Language: Ideologies, Polities and Identities*, ed. P. V. Kroskrity, 35–84. Santa Fe: School of American Research Press, 2000.
Jackson, A. *Home Rule: An Irish History 1800–2000.* Oxford: Oxford University Press, 2004.
Jackson, J. P., and N. Weidman. *Race, Racism and Science: Social Impact and Interaction.* New Brunswick, NJ: Rutgers University Press, 2006.
Jackson, R. *The Ascent of John Tyndall: Victorian Scientist, Mountaineer and Public Intellectual.* Oxford: Oxford University Press, 2018.
Jackson-Houlston, C. M. "'Queen Lilies'? The Interpenetration of Scientific, Religious and Gender Discourses in Victorian Representations of Plants." *Journal of Victorian Culture* 11, no. 1 (Spring 2006): 84–110.

Jarrell, R. "Colonialism and the Truncation of Science in Ireland and French Canada during the Nineteenth Century." *Journal of the History of Canadian Science Technology and Medicine* 5, no. 2 (18) (1981): 140–57.
Jarrell, R. "The Department of Science and Art and the Control of Irish Science, 1853–1905." In Bowler and Whyte, *Science and Society in Ireland*, 340–47.
Jefferson, T. "The Declaration of Independence" (1776), accessed June 27, 2019, https://www.archives.gov/founding-docs/declaration-transcript.
Johnson, N. "Botanical Travel, Climate and David Moore's Moral Geographies of Europe." *Journal of Historical Geography* 30, no. 1 (2013): 1–11.
Johnson, N. "Grand Design(er)s: David Moore, Natural Theology and the Royal Botanic Gardens in Glasnevin, Dublin." *Cultural Geographies* 14, no. 1 (2007): 29–55.
Johnson, N. "Making Space: Gaeltacht Policy and the Politics of Identity." In *In Search of Ireland: A Cultural Geography*, ed. B. Graham, 174–92. London: Routledge, 1997.
Johnson, N. *Nature Displaced, Nature Displayed: Order and Beauty in Botanical Gardens*. London: I. B. Tauris, 2011.
Jones, G. "Catholicism, Nationalism and Science." *Irish Review* 20 (1997): 47–61.
Jones, G. "Contested Territories: Alfred Cort Haddon, Progressive Evolutionism and Ireland." *History of European Ideas* 24 (1998): 195–211.
Jones, G. "Darwinism in Ireland." In Attis and Mollan, *Science and Irish Culture*, 115–39.
Jones, G. "Scientists against Home Rule." In *Defenders of the Union: A Survey of British and Irish Unionism since 1801*, ed. A. O'Day and D. G. Boyce, 188–208. London: Routledge, 2001.
Jones, G. *Social Darwinism and English Thought*. Brighton: Harvester, 1980.
Jowett, B. *The Politics of Aristotle*. Oxford: Clarendon Press, 1885.
Kargon, R. *Science in Victorian Manchester: Enterprise and Expertise*. Piscataway, NJ: Transaction, 2009.
Keith, A. *The Antiquity of Man*. London: Williams and Norgate, 1915.
Kelham, B. "The Royal College of Science for Ireland (1867–1926)." *Studies: An Irish Quarterly Review* 56, no. 223 (1967): 297–309.
Kelly, L. "'Fascinating Scalpel-Wielders and Fair Dissectors': Women's Experience of Irish Medical Education, c. 1880s–1920s." *Medical History* 54, no. 4 (October 2010): 495–516.
King, C. J. "Charles Moore (1820–1905)." *Australian Dictionary of Biography*, accessed May 30, 2018, http://adb.anu.edu.au/biography/moore-charles-4229.
Kingsley, F. *Charles Kingsley: His Letters and Memories of His Life*. London: Macmillan, 1895.
Kipling, R. *Traffics and Discoveries*. New York: Doubleday, Page and Company, 1904.
Knight, D. "Romanticism and the Sciences." In Cunningham and Jardine, *Romanticism and the Sciences*, 13–25.
Koertge, N. *New Dictionary of Scientific Biography*. 2nd ed. 8 vols. New York: Scribner, 2007.
Kohler, R. E. "A Generalist's Vision." *Isis* 96, no. 2 (2005): 224–29.
Kohler, R. E. "Lab History: Reflections." Focus: Laboratory History, *Isis* 99, no. 4 (2008): 761–68.

Kohler, R. E. "Labscapes: Naturalizing the Lab." *Journal of the History of Science* 40 (2002): 473–501.

Kohler, R. E. "Place and Practice in Field Biology." *Journal of the History of Science* 40 (2002): 189–210.

Korf, H. W., and H. Wicht. "The Public Cadaver." 428, no. 6985 *Nature* (April 2004): 805.

Kragh, H. "The New Physics." In *The Fin-de-Siècle World*, ed. M. Saler, 441–56. London: Routledge, 2014.

Kropotkin, P. *Mutual Aid: A Factor in Evolution.* London: Heineman, 1902.

Kuklick, H. *The Savage Within: The Social History of British Anthropology 1885–1945.* Cambridge: Cambridge University Press, 1991.

Kuklick, H., and R. E. Kohler. "Introduction." In "Science in the Field," ed. H. Kuklick and R. E. Kohler. *Osiris* 11, Second series (1996): 1–14.

Kuper, A. "Ancestors: Henry Maine and the Constitution of Primitive Society." *History and Anthropology* 1, no. 2 (1985): 265–86.

Lahire, B. *l'Homme Pluriel.* Paris: Nathan, 1998.

Laird, H. "India and the Translation of the Irish Brehon Laws." In Foley and O'Connor, *Ireland and India*, 3–11.

Larmor, J. *Aether and Matter.* Cambridge: Cambridge University Press, 1900.

Larmor, J. *The Scientific Writings of George Francis Fitzgerald . . . Collected and Edited with a Historical Introduction by J. Larmor.* Dublin: Hodges, Figgis, 1902.

Latour, B. *Pandora's Hope: Essays on the Reality of Science Studies.* Cambridge, MA: Harvard University Press, 1999.

Latour, B. *Science in Action: How to Follow Scientists and Engineers through Society.* Cambridge, MA: Harvard University Press, 1987.

Lawless, E. "North Clare-Leaves from a Diary." *Nineteenth Century* 46 (1899): 603–12.

Lawrence, C., and S. Shapin, eds. *Science Incarnate: Historical Embodiments of Natural Knowledge.* Chicago: University of Chicago Press, 1998.

Le Roux, L. N. *L'Irlande Militante: Le Vie de Patrick Pearse.* Rennes: Imprimerie Commerciale de Bretagne, 1932.

Leaney, E. "Missionaries of Science: Provincial Lectures in Nineteenth-Century Ireland." *Irish Historical Studies* 34 (2005): 266–88.

Lenoir, R., "Scientific Habitus: Pierre Bourdieu and the Collective Intellectual." *Theory, Culture, Society* 23 (2006): 25–43.

Levin, M. R., S. Forgan, M. Hessler, R. Kargon, and M. Low, eds. *Urban Modernity: Cultural Innovation in the Second Industrial Revolution.* Cambridge, MA: MIT Press, 2010.

Levy, D. M., and S. J. Peart, "The Secret History of the Dismal Science." Part 1. Economics. *The Online Library of Economics and Liberty (Econlib)*, accessed June 25, 2019, https://www.econlib.org/library/Columns/LevyPeartdismal.html.

Lightman, B., ed. *A Companion to the History of Science.* Chichester: Wiley-Blackwell, 2016.

Lightman, B. *The Origins of Agnosticism: Victorian Unbelief and the Limits of Knowledge.* Baltimore: Johns Hopkins University Press, 1987.

Lightman, B. "Refashioning the Spaces of London Science: Elite Epistemes in the Nineteenth Century." In Livingstone and Withers, *Geographies of Nineteenth-Century Science*, 25–50.

Lightman, B., ed. *Victorian Science in Context.* Chicago: University of Chicago Press, 1997.

Livingstone, D. N. *Adam's Ancestors: Race, Religion and the Politics of Human Origins.* Baltimore: Johns Hopkins University Press, 2008.

Livingstone, D. N. "Darwin in Belfast." In Foster, *Nature in Ireland*, 387–408.

Livingstone, D. N. "Darwinism and Calvinism: The Belfast-Princeton Connection." *Isis* 83, no. 3 (1992): 408–28.

Livingstone, D. N. *Dealing with Darwin: Place, Politics and Rhetoric in Religious Engagements with Evolution.* Baltimore: Johns Hopkins University Press, 2014.

Livingstone, D. N. "Natural Theology and Neo-Lamarkism: The Changing Context of Nineteenth-Century Geography in the United States and Great Britain." *Annals of the Association of American Geographers* 74, no. 1 (1984): 9–28.

Livingstone, D. N. *Putting Science in Its Place: Geographies of Scientific Knowledge.* Chicago: University of Chicago Press, 2003.

Livingstone, D. N. "Science, Region and Religion: The Reception of Darwinism in Princeton, Belfast and Edinburgh." In Numbers and Stenhouse, *Disseminating Darwinism*, 7–38.

Livingstone, D. N. "Science, Site and Speech: Scientific Knowledge and the Spaces of Rhetoric." *History of the Human Sciences* 20 (2007): 71–98.

Livingstone, D. N. "Science, Text and Space: Thoughts on the Geography of Reading." *Transactions of the Institute of British Geographers* 30 (2005): 391–401.

Livingstone, D. N. "Spaces of Knowledge: Contributions towards a Historical Geography of Science." *Environment and Planning D: Society and Space* 13 (1995): 5–34.

Livingstone, D. N., and C. W. Withers. *Geography and Enlightenment.* Chicago: University of Chicago Press, 1999.

Livingstone, D. N., D. G. Hart, and M. A. Noll, eds. *Evangelicals and Science in Historical Perspective.* Oxford: Oxford University Press, 1999.

Livingstone, D. N., and C. W. Withers, eds. *Geographies of Nineteenth-Century Science.* Chicago: University of Chicago Press, 2011.

Livingstone, D. N., and C. W. Withers. "Thinking Geographically about Nineteenth-Century Science." In Livingstone and Withers, *Geographies of Nineteenth-Century Science*, 1–21.

Lodge, O. *The Ether of Space.* New York: Harper Brothers, 1909.

Lodge, O. *Past Years: An Autobiography.* London: Hodder and Stoughton, 1931.

Lodge, O. "Scope and Tendencies of Physics." In *The 19th Century: A Review of Progress*, ed. A. Sedgwick et al., 348–57. London: G. P. Putnam's Sons, 1901.

Lubbock, J. *The Origin of Civilisation and the Primitive Condition of Man.* London: D. Appleton, 1871.

Lubbock, J. *Prehistoric Times: As Illustrated by Ancient Remains, and the Manners and Customs of Modern Savages.* New York: D. Appleton, 1865.

Luckhurst, R. *The Invention of Telepathy.* Oxford: Oxford University Press, 2002.

Lyons, J. B. "Medicine and Literature in Ireland." *Journal of the Irish Colleges of Physicians and Surgeons* 3, no. 1 (1973): 3–9.
Lyons, J. B. "George Sigerson: Charcot's Translator." *Journal of the History of the Neurosciences* 6, no. 1 (1997): 50–60.
Lyons, J. B. "Sigerson, George." In McGuire and Quinn, *Dictionary of Irish Biography*, 945–46.
Lysaght, S. *Robert Lloyd Praeger: The Life of a Naturalist.* Dublin: Four Courts Press, 1998.
Lysaght, S. "Themes in the Irish History of Science." *Irish Review* 19 (1996): 87–97.
Macalister, A. "The Bearing of Anomalous Anatomy on the Evolution Theory of the Origin of Man." *Dublin Journal of Medical Science* 53, no. 3 (March 1872): 192–98.
Macalister, A. *Evolution in Church History.* Dublin: Hodges, Figgis, 1882.
Macalister, A. *An Introduction to Animal Morphology and Systematic Zoology. Part 1. Invertebrata.* London: Longmans Green, 1876.
Macalister, A. *An Introduction to the Systematic Zoology and Morphology of Vertebrate Animals.* Dublin: Hodges, Foster and Figgis, 1878.
Macalister, A. "Man Physiologically Considered." *Present Day Tracts, on Subjects of Christian Evidence, Doctrine, and Morals* 7, no. 39 (1886): 1–48.
Macalister, A. *Memoir of James Macartney.* London: Hodder and Stoughton, 1901.
Macalister, A. "Review of *The Descent of Man.*" *Dublin Quarterly Journal of Medical Science* 52, no. 103 (1871): 133–52.
Macalister, A. "Review of Works on Life and Organisation." *Dublin Quarterly Journal of Medical Science* 50 (1870): 112–32.
Macalister, A. *Some Morphological Lessons Taught by Human Variation.* Oxford: H. Frowde, 1897.
MacCartney, D. "MacNeill and Irish Ireland." In Martin and Byrne, *Scholar Revolutionary*, 77–98.
MacDonald, H. "Procuring Corpses: The English Anatomy Inspectorate, 1842 to 1858." *Medical History* 53 (2009): 379–96.
MacDonald, W. *Reminiscences of a Maynooth Professor.* London: Mercier Press, 1925.
Mac Giolla Chriost, D. *The Irish Language in Ireland: From Goidel to Globalisation.* Routledge Studies in Linguistics. London: Routledge, 2005.
MacNeill, E. *Early Irish Laws and Institutions.* Dublin: Burns Oates and Washbourne, 1934.
MacNeill, E. "Ireland and Wales in the History of Jurisprudence." *Studies* 16 (1927): 254–58, 605–15.
MacNeill, E. "The Irish Law of Dynastic Succession." *Studies* 8 (1919): 367–82, 640–53.
MacNeill, E. "The Irish Nation and Irish Culture." *Ireland American Review* 1 (1938): 77–87.
MacNeill, E. "Irish Phonetics." *Claidheamh Solais* 5, no. 3 (1904): 8–9.
MacNeill, E. "Method in Teaching Irish." *Claidheamh Soluis* 7 (December 1905): 9.
MacNeill, E. "The Modern Standard of Irish." *Gaelic Journal* 8 (1898): 180–81.
MacNeill, E. "Ni h-Iad Na Mna Breaghtha Ghnios an Brochan Acht Min." *Gaelic Journal* 15 (1905): 17–19.

MacNeill, E. "On the Irish Infinitive." *Gaelic Journal* 4 (1892): 116–18.
MacNeill, E. "Our Whig Inheritance." *Ireland To-Day* 1, no. 6 (November 1936): 63–66.
MacNeill, E. "Past, Present and Future." *Gaelic Journal* 8 (1897): 2–4, 26–27.
MacNeill, E. *Phases of Irish History*. Dublin: M. H. Gill, 1920.
MacNeill, E. "Race and Nationality." *Claidheamh Soluis* 9 (October 1907): 7–8.
MacNeill, E. "Some Irish Law Documents." *Gaelic Journal* 8 (1897): 74–75, 86–90, 99–100, 114–15.
MacNeill, E. "Some Notes of Our National Literature." *Claidheamh Soluis* 5 (October 1907): 149.
MacNeill, E. "Some Points in Spelling." *Gaelic Journal* 9 (1898): 262–63, 271–72, 312, 347–48.
MacNeill, E. "The Study of Irish." *Gaelic Journal* 6 (1896): 159, 171–72, 188.
MacNeill, E. "A Vindication of Our Ancient Irish Culture." *Gaelic Journal* 8 (October 1897): 80–81.
MacVeagh, J. "Religious Intolerance under Home Rule: Some Opinions of Leading Irish Protestants." *Irish Press Agency* 22 (1911): 1–62.
Maguire, M. "A Socio-Economic Analysis of the Dublin Protestant Working Class, 1870–1926." *Irish Economic and Social History* 20, no. 1 (1993): 35–61.
Malcolm, E., and G. Jones. *Medicine, Disease and the State in Ireland 1650–1940*. Cork: Cork University Press, 1999.
Margulis, L. *Symbiotic Planet: A New Look at Evolution*. New York: Basic Books, 2008.
Marshall, J. "The Archaeology of the Abstract Body: Parascientific Discourse and the Legacy of Dr. J. M. Charcot." *French History and Civilisation* 3 (2009): 92–111.
Martin, F. X., and F. J. Byrne, eds. *The Scholar Revolutionary: Eoin MacNeill 1867–1945, and the Making of the New Ireland*. Dublin: Irish University Press, 1973.
Mathews P. J. *Revival: The Abbey Theatre, Sinn Féin, the Gaelic League and the Co-operative Movement*. Cork: Cork University Press (in association with Field Day), 2003.
Matless, D. "Original Theories: Science and the Currency of the Local." *Cultural Geographies* 10 (2003): 354–78.
Maume, P., and T. C. Edwards. "Eoin MacNeill." In McGuire and Quinn *Dictionary of Irish Biography*, 150–54.
Mauss, M. "Une catégorie de l'esprit humain: La notion de personne. Celle de 'moi.' Un plan de travail." *Journal of the Royal Anthropological Institute* 68 (1938): 263–81.
McCorristine, S. "Science and Nation." *Dublin Review of Books*, accessed May 29, 2018, http://drb.ie/essays/science-and-nation.
McCorristine, S. "William Fletcher Barrett, Spiritualism and Psychical Research in Edwardian Dublin." *Estudios Irlandeses* 6 (2011): 39–53.
McCracken, E. "The Botanic Gardens, Trinity College Dublin." *Garden History* 7, no. 1 (1979): 86–90.
McDowell, R. B. *The Irish Administration, 1801–1914*. London: Routledge and Kegan Paul, 1964.
McGeachie, J. "Science, Politics and the Irish Literary Revival: Reassessing 'Dr Sigerson' as Polymath and Public Intellectual." In *Cultures of Care in Irish Medical His-*

tory 1750–1970, ed. C. Cox and M. Luddy, 113–33. Basingstoke: Palgrave Macmillan, 2010.

McGilloway, K. *George Sigerson: Poet, Patriot, Scientist and Scholar.* Belfast: Star Uladh, 2011.

McGuire, J., and J. Quinn, eds. *Dictionary of Irish Biography*, vol. 8. Dublin and Cambridge: Royal Irish Academy and Cambridge University Press, 2009.

McLeod, R. "On Science and Colonialism." In Bowler and Whyte, *Science and Society in Ireland*, 1–18.

McMahon, T. *Grand Opportunity: The Gaelic Revival and Irish Society 1893–1910.* New York: Syracuse University Press, 2008.

Meenan, F. O. C. "The Catholic University School of Medicine 1860–1880." *Studies: An Irish Quarterly Review* 278/279, no. 70 (1981): 135–44.

Meijer, M. C. *Race and Aesthetics in the Anthropology of Petrus Camper (1722–1789).* Studies in the History of Ideas in the Low Countries. Amsterdam: Rodopi, 1999.

Messenger, J. "Literary vs. Scientific Interpretations of Cultural Reality in the Aran Islands of Eire." *Ethnohistory* 11, no. 1 (1964): 41–55.

Mitchell, M. E. "Reporting the Irish Lichen Fauna 1850–1900." *Glasra* 3 (1998): 155–60.

Mivart, St. G. J. *Man and Apes: An Exposition of Structural Resemblances and Differences Bearing upon Questions of Affinity and Origin.* New York: Appleton, 1874.

Mivart, St. G. J. *On the Genesis of Species.* New York: D. Appleton, 1871.

Mollan, C. "George Francis Fitzgerald, 1851–1901." In Mollan, *It's Part of What We Are*, 1199–224.

Mollan, C., ed. *It's Part of What We Are: Some Irish Contributions to the Development of the Chemical and Physical Sciences.* Science and Irish Culture 3. Vols. 1 and 2. Dublin: Royal Dublin Society, 2007.

Mollan, C. "James McCullagh (1809–1847)." In Mollan, *It's Part of What We Are*, 667–85.

Mollan, C. "William Fletcher Barrett, 1844–1925." In Mollan, *It's Part of What We Are*, 1112–31.

Moore, D. "Deconstructing Darwinism: The Politics of Evolution in the 1860s." *Journal of the History of Biology* 24, no. 3 (1991): 353–408.

Moore, D. *Hand-book for the Botanic Gardens of the Royal Dublin Society, Glasnevin.* Dublin: William Leckie, 1850.

Moore, D. "List of Hepaticae Which Are Found in the Counties of Dublin and Wicklow with Their Principal Localities." *Scientific Proceedings of the Royal Dublin Society* 1 (1878): 250–58.

Moore, D. "List of the Mosses of the Counties of Dublin and Wicklow with Their Principal Localities." *Scientific Proceedings of the Royal Dublin Society* 1 (1878): 228–49.

Moore, D. "On Growing Orchids from Seeds." *Gardeners' Chronicle* 35, no. 1 (September 1849): 549.

Moore, D. "The Structure &c. of Plants." In Porter, *Science and Revelation*, 1–32.

Moore, D., and A. G. More "Catalogue of the Flowering Plants and Ferns of Dublin and Wicklow." *Scientific Proceedings of the Royal Dublin Society* 1 (1878): 190–227.

Moore, D., and A. G. More. *Contributions towards a Cybele Hibernica: Being Outlines of the Geographical Distribution of Plants in Ireland.* Dublin: Hodges Smith, 1866.

Moore, F. W. "Some Reminiscences." *New Flora and Sylva* 12 (1940): 180–87.

Moran, D. P. "The Battle of Two Civilisations." In *Ideals in Ireland*, ed. Lady Gregory, 25–43. London: At the Unicorn, 1901.

Moran, D. P. *The Philosophy of Irish Ireland*, 1905. Reprinted with an introduction by Patrick Maume. Dublin: University College Dublin Press, 2007.

Morton, S. *Crania Americana.* London: Simpkin Marshall, 1839.

Morus, I. *When Physics Became King.* Chicago: University of Chicago Press, 2005.

Moseley, R. "Tadpoles and Frogs: Some Aspects of the Professionalisation of British Physics, 1870–1939." *Social Studies of Science* 7 (1977): 423–46.

Müller, F. M. *Natural Religion: The Gifford Lectures Delivered before the University of Glasgow in 1888.* London, 1889.

Müller, F. M. *Three Lectures on the Science of Language.* London: Longmans Green, 1891.

Mullholland, J. "James Macpherson's Ossian Poems: Oral Traditions and Invention of Voice." *Oral Tradition* 24, no. 2 (2009): 393–414.

Murray, R. H. *Archbishop Bernard: Professor, Prelate, Provost.* London: Society for Promoting Christian Knowledge, 1931.

Naylor, S. "Historical Geography: Knowledge in Place and on the Move." *Progress in Human Geography* 29, no. 5 (2005): 626–34.

Naylor, S. "Introduction: Historical Geographies of Science—Places, Contexts, Cartographies." *British Journal for the History of Science* 38, no. 1 (March 2005): 1–12.

Naylor, S. "Place in the History of Knowledge." Paper presented at the Eleventh Annual European Conference on the History of Economics, "Economics and Place." University of Paris, Nanterre, October 20–21, 2006.

Naylor, S. *Regionalising Science: Placing Knowledges in Victorian England.* London: Pickering and Chatto, 2010.

Nelson, C. "National Botanic Gardens Glasnevin: Retrospect and Prospect." *Curtis's Botanical Magazine* 12, no. 4 (1995): 182–85.

Nelson, C., and E. M. McCracken. *The Brightest Jewel: A History of the National Botanic Gardens Glasnevin, Dublin.* Kilkenny: Boethius Press, 1987.

Nelson, C., and M. Seward. "Charles Darwin's Correspondence with David Moore of Glasnevin on Insectivorous Plants and Potatoes." *Biological Journal of the Linnean Society* 15, no. 2 (1981): 157–63.

Neswald, E. "Science and Sociability in Nineteenth-Century Provincial Ireland: The Galway Mechanics' Institute." *British Journal for the History of Science* 39, no. 4 (2006): 503–34.

Nicholas, Rev. W. "Why Are the Methodists of Ireland Opposed to Home Rule?" *Irish Unionist Alliance* 1 (1893): 336–60.

Noakes, R. "'The Bridge Which Is between Physical and Psychical Research': William Fletcher Barrett, Sensitive Flames, and Spiritualism." *History of Science* 42 (2004): 419–64.

Noakes, R. "Ethers, Religion and Politics in Late Victorian Physics: Beyond the Wynne Thesis." *History of Science* 43 (2005): 415–55.

Nordau, M. *Degeneration.* New York: D. Appleton, 1895.
Numbers, R. L., and J. Stenhouse, eds. *Disseminating Darwinism: The Role of Place, Race, Religion and Gender.* Cambridge: Cambridge University Press, 1999.
Nye, M. J. "Scientific Biography: History of Science by Another Means?" *Isis* 97, no. 2 (2006): 322–29.
Nyhart, L. *Biology Takes Form: Animal Morphology and the German Universities, 1800–1900.* Chicago: University of Chicago Press, 1995.
Nyhart, L. "The Historiography of Science." In Lightman, *Companion to the History of Science*, 7–22.
O'Brien, S. "Thomas Acton: A Centennial Exhibition at Kilmacurragh." *International Dendrology Society Yearbook* (2008): 159–69.
O'Casey, S. *Drums under the Window.* London: Macmillan, 1946.
O'Conchubhair, B. *Fin de Siècle na Gaeilge: Darwin, Athbheochan agus Smaointeoireacht na hEorpa.* Dublin: Conradh na Gaeilge, 2009.
O'Cuiv, B. "MacNeill and the Irish Language." In Martin and Byrne, *Scholar Revolutionary*, 1–10.
O'Kelly, F. *A Guide to Early Irish Law.* Dublin: Institute for Advanced Studies, 2005.
O'Leary, D. *Roman Catholicism and Modern Science: A History.* New York: Continuum International, 2006.
O'Sullivan, T. "Locating Dublin in the Late Nineteenth-Century Ether." In Hochadel and Nieto-Galan, *Urban Histories of Science*, 102–22.
O'Sullivan, T. "The Perception of Place and the Origins of Handedness Debate: Towards a Cognitive Cartography of Science in late-Victorian Dublin." *Endeavour* 39, nos. 3–4 (2015): 139–46.
O'Tuama, S. *The Gaelic League Idea.* Cork: Mercier, 1972.
Ogle, W. "On Dextral Pre-eminence." *Transactions of the Medical and Chirurgical Society of London* 54 (1871): 279–301.
Olry, R. "Body Snatchers: The Hidden Side of the History of Anatomy." *Journal of the International Society of Plastination* 14, no. 2 (1999): 6–9.
Ophir, A., and S. Shapin. "The Place of Knowledge: A Methodological Survey." *Science in Context* 4 (1991): 3–21.
Ordnance Survey of Ireland. *Ordnance Survey of the County of Londonderry.* Published for H. M. Government, Hodges and Smith, 1837.
Ospovat, D. *The Development of Darwin's Theory: Natural History, Natural Theology and Natural Selection, 1838–1859.* Cambridge: Cambridge University Press, 1981.
Ospovat, D. "Perfect Adaptation and Teleological Explanation." *Studies in the History of Biology* 2 (1978): 33–57.
Ostoff, H., and K. Brugmann. *The Morphological Investigation in the Sphere of the Indo-European Languages: A Reader in Nineteenth-Century Historical Indo-European Languages.* 1878, vol. 1. Edited and translated by W. Lehmann. Bloomington: Indiana University Press, 1967.
Outram, D. "New Spaces in Natural History." In *Cultures of Natural History*, ed. N. Jardine, J. Secord, and E. Spary, 249–66. Cambridge: Cambridge University Press, 1996.

Owen, R. *On the Anatomy of Vertebrates.* London: Longmans Green, 1866.

Owen, R. *On the Archetype and Homologies of the Vertebrate Skeleton.* London: Richard and John E. Taylor, 1847.

Owen, R. *On the Nature of Limbs.* London: John Van Voorst, Paternoster Row, 1849.

Paley, W. *Natural Theology: Or, Evidences of the Existence and Attributes of the Deity.* 12th ed. London: J. Faulder, 1809.

Patten, E. "Ireland's 'Two Cultures' Debate: Victorian Science and the Literary Revival." Special Issue, "New Perspectives on the Irish Literary Revival." *Irish University Review* 31, no. 1 (Spring–Summer 2003): 1–13.

Pick, D. *Faces of Degeneration: A European Disorder c. 1848–1891.* Cambridge: Cambridge University Press, 1989.

Plunkett, H. *Ireland in the New Century.* London: John Murray, 1904.

Porter, J. L. *Science and Revelation: A Series of Lectures in Reply to the Theories of Tyndall, Huxley, Darwin and Spencer Etc.* Belfast: William Mullan, 1875.

Porter, T. M. "Is the Life of the Scientist a Scientific Unit?" *Isis* 97, no. 2 (2006): 314–21.

Power, A. *Conversations with James Joyce.* New York: Harper and Row, 1974.

Poynting, J. H. *Collected Scientific Papers.* London: Cambridge University Press, 1920.

Praeger, L. *A Populous Solitude.* London: Methuen, 1941.

Preston, T. *The Theory of Light.* London: Macmillan, 1895.

Price, G. R., and J. B. Stewart. *To Heal the Scourge of Prejudice: The Life and Writings of Hosea Easton.* Amherst: University of Massachusetts Press, 1999.

Purser, M. "As We Saw It: The Context of the Life of George Francis Fitzgerald." *European Review* 15, no. 4 (2007): 523–29.

Pye-Smith, P. H. "Left-handedness." *Guy's Hospital Reports* 16 (1871): 141–46.

Quetelet, A. *Sur l'homme et le développement de ses facultés, ou essai de physique sociale.* Paris: Bachelier, 1835.

Quiggin, A. H. *Haddon the Head-Hunter: A Short Sketch of the Life of A. C. Haddon.* Cambridge: Cambridge University Press, 1942.

Radick, G. "Race and Language in the Darwinian Tradition (and What Darwin's Language Species Parallels Have to Do with It)." *Studies in Historical, Philological, Biological and Biomedical Sciences* 39 (2008): 359–70.

Radick, G. *The Simian Tongue: The Long Debate about Animal Language.* Chicago: University of Chicago Press, 2007.

Randall, L. *Warped Passages: Unravelling the Universe's Hidden Dimensions.* New York: Harper Collins, 2005.

Ray, J. *The Wisdom of God Manifested in the Works of Creation.* 7th ed. London: W. Innys, 1714.

Rees, M. *Our Cosmic Habitat: New Edition.* Princeton, NJ: Princeton University Press, 2017.

Renan, E. *La Poésie des races celtiques.* Paris: J. Claye et Cie, 1854.

Rich, P. B. *Race and Empire in British Politics.* Cambridge: Cambridge University Press, 1986.

Richards, J. "Introduction: Fragmented Lives." *Isis* 97, no. 2 (2006): 302–5.

Richards, R. "Darwin's Place in the History of Thought: A Re-Evaluation." *Proceedings of the National Academy of Sciences* 106, no. 1 (2009): 10056–60.

Richards, R. *The Romantic Conception of Life: Science and Philosophy in the Age of Goethe*. Chicago: University of Chicago Press, 2002.

Richardson, P. "The Consolation of Philology." *Modern Philology* 92, no. 1 (August 1994): 1–13.

Richardson, R. *Death, Dissection and the Destitute*. London: Routledge and Kegan Paul, 1988.

Riley, A. "Crisis, *Habitus* and Intellectual Trajectory." *European Journal of Social Sciences* 42, no. 129 (2004): 307–14.

Robertson, G. M. "Hypnotism at Paris and Nancy: Notes of a Visit." *British Journal of Psychiatry* 38, no. 163 (1892): 494–531.

Robins, R. H. *A Short History of Linguistics*. London: Longman, 1968.

Robinson, W. "Notes on Gardens XIX–XXII: Royal Botanic Gardens Glasnevin." *Gardeners' Chronicle* (1864): 988–89.

Rock, H., and D. Moore. *Cityscapes*. New York: Columbia University Press, 2001.

Rorty, A. *The Identities of Persons*. Berkeley: University of California Press, 1969.

Rothenberg, M., ed. "Joseph Henry Papers," accessed May 29, 2018, http://siarchives.si.edu/history/jhp/joseph08.htm.

Rudwick, M. *The Great Devonian Controversy: The Shaping of Scientific Knowledge among Gentlemanly Specialists*. Chicago: University of Chicago Press, 1985.

Rupke, N. "Afterword." In Livingstone and Withers, *Geographies of Nineteenth-Century Science*, 439–50.

Rupke, N. *Alexander von Humboldt: A Metabiography*. Bern: Peter Lang, 2008.

Rupke, N. "A Geography of Enlightenment: The Critical Reception of Alexander von Humbolt's Mexico Work." In Livingstone and Withers, *Geography and Enlightenment*, 319–45.

Rupke, N. *Richard Owen: Biology without Darwin*. Chicago: University of Chicago Press, 2009.

Rupke, N. *Richard Owen: Victorian Naturalist*. New Haven, CT: Yale University Press, 1994.

Rupke, N. "Richard Owen's Vertebrate Archetype." *Isis* 84, no. 2 (1993): 231–51.

Russell, E. R. *Form and Function: A Contribution to the History of Animal Morphology*. London: John Murray, 1916.

Ryan, J. "Eoin MacNeill, 1867–1945." *Studies: An Irish Quarterly Review* 34, no. 136 (December 1945): 433–48.

Said, E. W. *The World, the Text, and the Critic*. London: Vintage, 1991.

Sambursky, S. *The Physical World of Late Antiquity*. London: Routledge and Kegan Paul, 1962.

Sappol, M. *A Traffic of Dead Bodies: Anatomy and Embodied Social Identity in Nineteenth-Century America*. Princeton, NJ: Princeton University Press, 2002.

Schaffer, S. "The Matter of Ether." *History of Science* 20, no. 4 (December 1982): 297–303.

Schleicher A. "The Darwinian Theory and the Science of Language." In *Linguistics and Evolutionary Theory*, ed. K. Koerner, 1–73. Amsterdam Classics in Linguistics, 1800–1925. Amsterdam: John Benjamin, 1983.

Scobie, C. "The Long Road Home." *Observer*, June 28, 2009, accessed May 29, 2018, https://clairescobie.com/wp-content/uploads/2014/12/On-repatriation-of-indigenous-remains-June-09.pdf.

Seward, A. C., and F. Darwin, eds. *More Letters of Charles Darwin: A Record of His Work in a Series of Hitherto Unpublished Letters*, vol. 2. London: John Murray, 1903.

Secord, A. "Be What You Would Seem to Be: Samuel Smiles, Thomas Edward and the Making of a Working-Class Scientific Hero." *Science in Context* 16, no. 1/2 (2003): 147–73.

Secord, J. "Introduction" to R. Chambers, *Vestiges of the Natural History of Creation and Other Evolutionary Writings*, vii–xlviii. Chicago: University of Chicago Press, 1994.

Secord, J. "Knowledge in Transit." *Isis* 95 (2004): 654–72.

Secord, J. *Victorian Sensation: The Extraordinary Publication, Reception, and Secret Authorship of Vestiges of the Natural History of Creation*. Chicago: University of Chicago Press, 2000.

Shapin, S. "Here and Everywhere: Sociology of Scientific Knowledge." *Annual Review of Sociology* 21 (1995): 289–321.

Shapin, S. "Placing the View from Nowhere: Historical and Sociological Problems in the Location of Science." *Transactions of the Institute of British Geographers* 23, no. 1 (1998): 5–12.

Shortland, M., and R. Yeo. *Telling Lives in Science: Essays on Scientific Biography*. Cambridge: Cambridge University Press, 1996.

Shteir, A. B. "Gender and 'Modern' Botany in Victorian England." *Osiris* 12 (1997): 29–38.

Shute, D. K. "Racial Anatomical Peculiarities." *American Anthropologist* 9, no. 4 (April 1896): 123–32.

Sigerson, G. *The Bards of Gael and Gall*. London: T. Fisher Unwin, 1887.

Sigerson, G. "Consideration of the Structural and Acquisitional Elements in Dextral Pre-Eminence with Conclusions as to the Ambidexterity of Primeval Man." *Proceedings of the Royal Irish Academy*, series 2, science, 4 (1884): 38–50.

Sigerson, G. "An Examination into Recently Reported Phenomena in Connexion with Hysteron-Epilepsy and Cerebral Anaesthesia." *British Medical Journal* (February 1879): 181–84.

Sigerson, G. "Genesis and Evolution." *New Ireland Review* 1 (March–August 1894): 18–26.

Sigerson, G. "Influence of Solonoids on the Nervous System." *British Medical Journal* (April 1879): 620–21.

Sigerson, G. "Irish Literature: Its Origin, Environment and Influence." In Duffy, Sigerson, and Hyde, *Revival of Irish Literature*. 61–115.

Sigerson, G. *The Last Independent Parliament of Ireland*. Dublin: M. Gill, 1919.

Sigerson, G. "Micro-Atmospheric Researches: With 7 Plates." *Proceedings of The Royal Irish Academy Science* series 2, no. 1 (1869–1874): 13–30.

Sigerson, G. *Modern Ireland: Its Vital Questions, Secret Societies and Government; By an Ulsterman*. London: Longmans, Green, Reader and Dyer, 1868.

Sigerson, G. "On Alternate Paralysis." *Dublin Journal of Medical Science*, third series, 35, no. 74 (1878): 98–135.

Smith, C. *The Science of Energy, A Cultural History of Energy Physics in Victorian Britain.* London: Athlone Press, 1998.

Smith, C., and J. Agar. "Introduction: Making Space for Science." In Smith and Agar, *Making Space for Science*, 1–23.

Smith, C., and J. Agar, eds. *Making Space for Science: Territorial Themes in the Shaping of Knowledge.* Basingstoke: Macmillan, 1998.

Smith, C., and M. N. Wise. *Energy and Empire: A Biographical Study of Lord Kelvin.* Cambridge: Cambridge University Press, 1989.

Smith, G. E. "Alexander Macalister, 1844–1919." *Proceedings of the Royal Society of London*, Series B, 94 (1922–1923): xxxii–xxxix.

Smith, J. *Charles Darwin and Victorian Visual Culture.* Cambridge: Cambridge University Press, 2006.

Smith, J. "Grant Allen: Physiological Aesthetics and the Dissemination of Darwin's Botany." In *Science Serialized: Representations of the Sciences in Nineteenth-Century Periodicals*, ed. G. Cantor and S. Shuttleworth, 285–307. Cambridge, MA: MIT Press, 2004.

Smith, J. "Une fleure du mal? Swinburne's 'The Sundew' and Darwin's Insectivorous Plants." *Victorian Poetry* 41, no. 1 (Spring 2003): 131–50.

Soderquist, T. "Existential Projects and Existential Choice in Science: Science Biography as an Edifying Genre." In Shortland and Yeo, *Telling Lives in Science*, 45–85.

Söderqvist, T., ed. *The History and Poetics of Scientific Biography.* Science, Technology and Culture 1700–1945. Aldershot: Ashgate, 2007.

Söderqvist, T. "Introduction: A New Look at the Genre of Scientific Biography." In Söderqvist, *History and Poetics of Scientific Biography*, 1–17.

Soja, E. W. *Postmetropolis: Critical Studies of Cities and Regions.* London: Wiley Blackwell, 2000.

Spearman, D. *Samuel Haughton: Victorian Polymath.* Dublin: Royal Irish Academy, 2002.

Spencer, H. *Social Statics: or the Conditions Essential to Human Happiness, Specified, and the First of Them Developed.* New York: D. Appleton, 1872.

Spencer, L. "Chance, Circumstance and Folly: Richard Berry and the Plaster Anatomical Collection of the Harry Brookes Allen Museum of Anatomy and Pathology." *University of Melbourne Collections* 2 (July 2008): 1–8.

Stepan, N. *Picturing Tropical Nature.* London: Reaktion, 2001.

Stewart, B., and P. G. Tait. *The Unseen Universe, or Physical Speculations on a Future State.* London: Macmillan, 1875.

Stocking, G. "The Persistence of Polygenist Thought in Post-Darwinian Anthropology." In Stocking, *Race, Culture and Evolution*, 42–69.

Stocking, G., ed. *Race, Culture and Evolution: Essays in the History of Anthropology.* New York: Free Press, 1968.

Stocking, G. *Victorian Anthropology.* London: Simon and Schuster, 1991.

Taylor, C. *Sources of the Self: The Making of Modern Identity.* Cambridge, MA: Harvard University Press, 1989.

Terrall, M. "Biography as Cultural History of Science." *Isis* 97, no. 2 (2006): 306–13.

Thompson, A. "Daniel John Cunningham M.D. D.C. L. F.R.S." *Journal of Anatomy and Physiology* 44, no. 4 (1909 third series): 1–3.

Thompson, A. "The Influence of Posture on the Articular Surfaces of the Tibia and Astragalus in the Different Races of Men and the Higher Apes." *Journal of Anatomy and Physiology* 23 (1899): 616–39.

Thompson, D. W. *On Aristotle as a Biologist with a Prooemion on Herbert Spencer; Being the Herbert Spencer Lecture Delivered before the University of Oxford on February 14th, 1913.* Oxford: Clarendon Press, 1913.

Thomson, W. "On Vortex Atoms." *Proceedings of the Royal Society of Edinburgh* 6 (1867): 94–105.

Thrift, N. "Remembering the Technological Unconscious by Foregrounding Knowledges of Position." *Environment and Planning D, Society and Space* 22 (2004): 175–90.

Tierney, M. "What Did the Gaelic League Accomplish?" *Studies: An Irish Quarterly Review* 52, no. 208 (Winter 1963): 337–47.

Topinard, P. "Man in Nature." *Popular Science* 42, no. 29 (1893): 445–53.

Torchiana, D. *Backgrounds for Joyce's Dubliners.* Boston: Allen and Unwin, 1986.

Tresch, J. *The Romantic Machine: Utopian Science and Technology after Napoleon.* Chicago: University of Chicago Press, 2012.

Trouton, F. "Address of the President to the Section of Mathematical and Physical Science of the British Association for the Advancement of Science." *Science* 2 (October 19142: 457–64.

Turner, F. *Between Science and Religion: The Reaction to Scientific Naturalism in Late Victorian England.* New Haven, CT: Yale University Press, 1974.

Turner, F. *Contesting Cultural Authority: Essays in Victorian Intellectual Life.* Cambridge: Cambridge University Press, 1993.

Turpin, J. "John Hogan and the Catholic Religious Revival." *Maynooth Review* 5, no. 1 (1979): 64–70.

Tylor, E. B. *Primitive Culture: Researches into the Development of Mythology, Philosophy, Religion, Language, Art and Custom.* Boston: Estes and Lauriat, 1874.

Tylor, E. B. *Researches into the Early History of Mankind and the Development of Civilisation.* London: John Murray, 1865.

Tyndall, J. *Address Delivered before the British Association Assembled at Belfast, with Additions.* London: Longmans Green, 1874.

Tyndall, J. *Essays on the Use and Limit of the Imagination in Science.* London: Longmans, Green, 1870.

Tyndall, J. *Fragments of Science for Unscientific People: A Series of Detached Essays, Lectures and Reviews.* New York: D. Appleton, 1871.

Tyndall, J. "Science and Spirits." In Tyndall, *Fragments of Science*, 402–10.

Uitti, K. D. *Linguistics and Literary Theory.* Humanistic Scholarship in America. Englewood Cliffs, NJ: Prentice Hall,1969.

Urry, J., ed. *Before Social Anthropology: Essays on the History of British Anthropology.* New York: Harwood Academic, 1993.

Urry, J. "Englishmen, Celts and Iberians: The Ethnographic Survey of the United Kingdom, 1892–1924." In Urry, *Before Social Anthropology*, 83–102.

Urry, J. "From Zoology to Ethnology: A. C. Haddon's Conversion to Anthropology." In Urry, *Before Social Anthropology*, 61–83.

Urry, J. "'Notes and Queries on Anthropology,' and the Development of Field Methods in British Anthropology, 1870–1920." *Proceedings of the Royal Anthropological Institute of Great Britain and Ireland* 1972 (1972): 45–57.

von Humboldt, W. *"On Language": On the Diversity of Human Language Construction and Its Influence on the Mental Development of the Human Species.* Cambridge Texts in the History of Philosophy, ed. M. Losonsky. Cambridge: Cambridge University Press, 1999.

Waddell, J. *Foundation Myths.* Wicklow: Wordwell, 2005.

Warwick, A. "The Sturdy Protestants of Science: Larmor, Trouten and the Earth's Motion through the Ether." In *Scientific Practice: Theories and Stories of Doing Physics*, ed. J. Z. Buchwald, 300–343. Chicago: University of Chicago Press, 1995.

Weaire, D. *George Francis Fitzgerald.* Pöllauberg: Living Edition, 2009.

Weaire D., and M. Coey. "Mentor and Constant Friend: The Life of George Francis Fitzgerald 1851–1901." *European Review* 15, no. 4 (2007): 513–21.

Webb, B. *My Apprenticeship.* London: Longmans Green, 1926.

Webb, T. E. *Veil of Isis.* Dublin: Hodges, Figgis, 1885.

West, T. *Horace Plunkett: Co-operation and Politics.* Gerrards Cross: Colin Smythe, 1986.

Westfall, R. *The Life of Isaac Newton.* Cambridge: Cambridge University Press, 1993.

Westwood, S., and J. Williams. "Imagining Cities." In *Imagining Cities: Scripts, Signs, Memory*, ed. Westwood and Williams, 1–19. London: Routledge, 1997.

Whewell, W. *History of the Inductive Sciences from the Earliest to the Present Times.* London: W. Parker, 1837.

White, C. *An Account of the Regular Gradation in Man, and in Different Animals and Vegetables, and from the Former to the Latter.* London: C. Dilly, 1799.

Whittaker, E. T. *A History of the Theories of Aether and Electricity: From the Age of Descartes to the Close of the Nineteenth Century.* London: Longmans Green, 1910.

Whyte, N. "Science and Nationality in Edwardian Ireland." In Bowler and Whyte, *Science and Society in Ireland*, 49–67.

Whyte, N. *Science, Colonialism and Ireland.* Cork: Cork University Press, 1999.

Williams, J. "The Irish Astronomical Tract: A Case Study of Scientific Terminology in 14th Century Irish." M.Ph. thesis, Department of Celtic Studies, University of Sydney, 2002.

Wilson, D. B. *Kelvin and Stokes: A Comparative Study in Victorian Physics.* Bristol: Adam Hilger, 1987.

Withers, C. "The Geography of Scientific Knowledge." In *Göttingen and the Development of the Natural Sciences*, ed. N. Rupke, 9–18. Göttingen: Wallstein Verlag, 2002.

Withers, C. "Place and the 'Spatial Turn' in Geography and in History." *Journal of the History of Ideas* 4, no. 4 (2009): 637–58.

Wood, M. "A Thunderbolt of Special Wrath." PhD diss., Queens University Belfast, 2011.

Yam, T. W., J. Arditti, and K. M. Cameron, K. M. "'The Orchids Have Been a Splendid Sport': An Alternative Look at Charles Darwin's Contribution to Orchid Biology." *American Journal of Botany* 96, no. 12 (2009): 2128–54.

Yeats, W.B. *The Celtic Twilight.* London: A. H. Bullen, 1893.

Yeo, R. "William Whewell, Natural Theology and the Philosophy of Science in Mid-Nineteenth Century Britain." *Annals of Science* 36 (1979): 493–516.

Zeldin, T. *Conversation.* London: Harvill Press, 1998.

Index